MW00675911

TEST BANK
to accompany

CHEMISTRY
& Chemical Reactivity
Fourth Edition

KOTZ & TREICHEL

RONALD O. RAGSDALE
University of Utah

SAUNDERS GOLDEN SUNBURST SERIES
Saunders College Publishing
Harcourt Brace College Publishers

Fort Worth Philadelphia San Diego New York Orlando Austin
San Antonio Toronto Montreal London Sydney Tokyo

Portions of this work were published in previous editions.

Printed in the United States of America

ISBN 0-03-023804-8

Preface

This test bank contains printed tests to accompany Kotz and Treichel's CHEMISTRY & CHEMICAL REACTIVITY, 4th edition. It contains 1200 multiple-choice questions on perforated pages. Answers are at the end of each chapter's worth of questions.

This test bank is available in computerized versions on disk for Macintosh and IBM (DOS and Windows). The ExaMaster+ TM computerized test banks allow you to preview, select, edit, and add items and to tailor tests to your course or select items at random and to add graphics. For more information, please contact your Saunders sales representative.

We have made every effort to review this test bank for accuracy and consistency. If you have any comments or suggestions about this test bank, please address your correspondence to Chemistry Editor, Saunders College Publishing, 150 S. Independence Mall West, Philadelphia, PA 19106-3412.

Chapter 1
Matter and Measurement

1. Which physical state(s) of matter exhibit(s) the greatest change in volume with changes in temperature or pressure?

   1. Solid
   2. Liquid
   3. Gas

   a. 1 only b. 2 only c. 3 only d. 1 and 3 only e. 1, 2, and 3

2. The smallest particle of an element that retains the chemical properties of the element is a(n)

   a) atom b) molecule c) ion d) solid e) gas

3. According to the Kinetic Molecular Theory particles of a solid

   a) are bound in a regular array and do not move.
   b) float freely within an array occupying various positions relative to neighbors.
   c) have no relationship to the microscopic structure of the solid.
   d) vibrate back and forth but do not move past immediate neighbors.
   e) float freely in the inside but no not move on the surface.

4. In the gaseous state, particles

   a) move independently and randomly.
   b) have strong attractions for each other.
   c) fill the available space because of strong interactions.
   d) lose energy in time.
   e) gain energy in time.

5. What is the density of a metal if a 15.4 gram sample has a volume of 1.96 cm^3?

   a) 0.127 g/cm^3 b) 0.511 g/cm^3 c) 7.86 g/cm^3 d) 30.2 g/cm^3 e) 33.1 g/cm^3

6. What volume of a liquid having a density of 1.48 g/cm^3 is needed to supply 5.00 grams of the liquid?

   a) 0.296 cm^3 b) 1.48 cm^3 c) 2.26 cm^3 d) 3.38 cm^3 e) 7.40 cm^3

7. The density of aluminum is 2.70 g/cm^3. If a cube of aluminum weighs 13.5 grams, what is the length of the edge of the cube?

   a) 5.00 cm b) 1.71 cm c) 1.25 cm d) 0.312 cm e) 0.200 cm

8. The dimensions of a rectangular solid are 8.45 cm long, 4.33 cm wide and 2.85 cm high. If the density of the solid is 9.43 g/ cm^3, what is its mass?

a) 1.12 grams b) 11.1 grams c) 154 grams d) 896 grams e) 983 grams

9. A metal sample weighing 30.9232 grams was added to a graduated cylinder containing 23.26 mL of water. The volume of the water plus the sample was 24.85 mL. What is the density of the metal?

a) 2.0×10^1 g/mL b) 19.4 g/mL c) 19.45 g/mL d) 19.449 g/mL

e) 19.4486 g/mL

10. Which temperature change is the smallest?

a) 10°C to 20°C b) 10 K to 20°C c) 10 K to 20 K d) 10°F to 20°C

e) 10°F to 20°F

11. In comparing the Kelvin and Celsius temperature scales which one of the following statements is true?

a) The difference between the freezing and boiling points of water is 273°C.

b) The Kelvin scale has no negative numbers.

c) The scales have the same zero point

d) –273 K = 0°C.

e) 0 K = 273°C.

12. Approximately how many elements are found in nature?

a) 185 b) 110 c) 90 d) 60 e) 44

13. The number of significant digits in 0.30500 is

a) 1 b) 2 c) 3 d) 4 e) 5

14. The number of significant figures in 0.06060×10^{-5} is

a) 2 b) 3 c) 4 d) 5 e) 6

15. A student does a calculation using her calculator and the number 280.27163 is shown on the display. If there are actually three significant figures, how should she show the final answer?

a) 280 b) 280.3 c) 280.27 d) 2.80×10^{-2} e) 2.80×10^2

16. The number, three hundred and fifty thousand, written in scientific notation is best written as

a) 350 b) 3.5×10^6 c) 3.5×10^5 d) 3.50×10^5 e) 3.50×10^{-5}

17. The mass of a sample weighed on an electronic balance that is sensitive ± 0.3 mg is 1.2300 g. The number of significant figures in this measurement is

a) 1 b) 2 c) 3 d) 4 e) 5

18. The value seven hundred fifty million with four significant figures is written as

a) 7500×10^6 b) 7.500×10^8 c) 750×10^6 d) 750,000,000 e) 7500×10^8

19. Evaluate the expression (8.346 + 2.854) (1.2750) =

a) 14.28 b) 1.428×10^1 c) 1.428×10^2 d) 1.428 e) 1.4280×10^1

20. What is the numerical value of 1.5 cm – 7.222×10^{-1} cm?

a) 0.7778 cm b) 0.778 cm c) 0.78 cm d) 0.8 cm e) 7.072×10^{-1} cm

21. When the masses 0.0222 kg, 140.000 g and 5888 mg are added, the total should be reported with _____ significant figures.

a) 3 b) 4 c) 5 d) 6 e) 7

22. Four samples were weighed using three different balances. (All are as accurate as the precision below indicates.) The masses are 0.94 kg, 58.2 g, 1.55 g, and 250 mg. This total mass should be reported as

a) 1000.000 g. b) 1000.0 g. c) 1.000×10^3 g. d) 1.00×10^3 g. e) 1.0×10^3 g.

23. The percentage water is an unknown hydrate was determined by heating the sample and driving the water off the sample. Two independent measurements gave values of 19.564 and 21.731%. Its percentage should be reported as

a) 20.6475% b) 20.648% c) 20.65% d) 20.6% e) 21%.

24. The temperature of the room is 75°F. What is its temperature in Celsius degrees?

a) 24°C b) 27°C c) 30°C d) 43°C e) 59°C

25. Which one of the following elements is an alkali metal?

a) Al b) Ag c) Au d) Pb e) Rb

26. Which of the following lists gives the symbol of a transition metal, a nontransition metal, and a non-metal <u>in that order</u>?

a) Fe, Cl, Ca b) Na, Fe, Cl c) Fe, In, Ar d) Na, In, Cl e) Ca, Fe, Ar

27. The symbols for a metal, a non-metal and a noble gas in that order are

a) Ag, Ga, Xe b) Ce, Ge, Ne c) Ca, Sn, Kr d) Ba, P, Ar e) P, Pb, Kr

28. Which of the following elements is a non-metal?

a) Ca b) Cr c) Co d) Cl e) Cs

29. Which element is a gas at room temperature?

a) argon b) calcium c) sulfur d) lead e) neodymium

30. A good example of an ionic compound is

a) water b) sugar c) dry ice d) sodium chloride e) natural gas

31. When a pure solid substance was heated, a student obtained another solid and a gas, each of which was a pure substance. From this information which of the following statements is ALWAYS a correct conclusion?

a) The original solid is not an element.
b) Both products are elements.
c) The original solid is a compound and the gas is an element.
d) The original solid is an element and the gas is a compound.
e) Both products are compounds.

32. Which of the following is a chemical property of water?

a) its density is 1.000 g/cm^3 at 4°C.
b) its melting point is 0°C
c) it forms bubbles when calcium is added
d) it causes light rays to bend
e) its heat of fusion is 6020 J/mol

33. Which of the following is an example of a chemical change?

a) water boiling b) ice melting c) natural gas burning d) iodine vaporizing
e) dry ice subliming

34. Classify each observation as a physical or a chemical property and tally them.

Observation 1: Bubbles form on a piece of metal when it is dropped into acid.

Observation 2: The color of a crystalline substance is yellow.

Observation 3: A shiny metal melts at 650°C.

Observation 4: The density of a solution is 1.84 g/cm^3

a) 2 chemical properties and 2 physical properties

b) 3 chemical properties and 1 physical properties

c) 1 chemical properties and 3 physical properties

d) 4 chemical properties

d) 4 physical properties

35. To convert a value in kilograms to centigrams one should

a) multiply by 10^5 b) multiply by 10^3 c) multiply by 10^{-3} d) divide by 10^5 e) divide by 10^{-1}

36. How many inches are in a length of 404 mm?

a) 15.9 in b) 103 in c) 1.026×10^2 in d) 159 in e) 1026 in

37. How many cm^2 are in an area of 4.21 in^2?

a) 10.7 cm^2 b) 114 cm^2 c) 27.2 cm^2 d) 1.66 cm^2 e) 1.14 cm^2

38. A wavelength of 17 nanometers is equal to _____ kilometers.

a) 17×10^9 km b) 17×10^{-9} km c) 17×10^{12} km d) 17×10^{-12} km

e) 1.7×10^{-10} km

39. A pressure of 1.00 lbs/in^2 is equal to _____.

a) $(454\ g)^2 \div 2.54$ cm b) $454\ g \div (2.54\ cm)^2$ c) $(454\ g \div 2.54\ cm)^2$

d) $2.54cm^2 \div 454$ g e) $[454\ g \div (2.54\ cm)^2]^2$

40. When the prefix micro (μ) is used in the metric system, a fundamental unit of measurement is multiplied by a factor of

a) 10^{-9} b) 10^{-6} c) 10^{-3} d) 10^3 e) 10^9

41. A rectangular piece of aluminum foil is measured as 2.00 cm long and 3.45 cm wide. What is the area of the sheet?

a) 7 cm^2 b) 6.9 cm^2 c) 6.90 cm^2 d) 6.9000 cm^2 e) 7.0×10^2 cm^2

42. Of the masses 86.30 g, 0.0863 kg and 8.630×10^5 mg , which (if any) is the largest?

a) 86.30 g b) 0.0863 kg c) 8.630×10^5 mg d) they are the same

e) two are the same, one is smaller

43. Which of the three statements regarding compounds is (are) true?

1. A compound has different properties than the elements of which it is composed.
2. A compound has a definite percentage composition by mass of its combining elements.
3. A compound must be composed of molecules.

a) 1 only b) 2 only c) 3 only d) 1 and 2 only e) 1, 2, and 3

44. The element chlorine is obtained for commercial use by the following method:

a) Isolation from gas pockets in the earth's crust.

b) Separation from air by a high pressure technique.

c) Filtration of brine (NaCl) solutions.

d) Electrolysis of aqueous NaCl solutions.

e) Mixing sulfur and argon in equal quantities.

45. Consider a brass alloy which contains 66% copper and 34% zinc. How many g of zinc are present in a 125 kg sample of the alloy?

a) 2.4 g b) 42 g c) 83 g d) 2.4×10^4 g e) 4.2×10^4 g

46. If a 4.50 g sample of a Zn-Al-Cu alloy contains 2.45 g Zn and 1.34 g Al, what is the % composition of Cu?

a) 3.8 % b) 7.1% c) 15.8% d) 71.0% e) 84.2%

47. How many kilograms of tin are in a 3.25 lb. roll of tin foil which is 92.0% tin and 8.00% zinc? (1 kg = 2.20 lbs.)

a) 1.36 kg b) 1.48 kg c) 6.78 kg d) 7.15 kg e) 11.8 kg

48. The density of a sodium sulfate solution is 1.07 g/cm^3. The solution is 8.00% sodium sulfate by mass. How many cm^3 of the solution are needed to supply 4.28 g of sodium sulfate?

a) 30.0 cm^3 b) 35.0 cm^3 c) 40.0 cm^3 d) 45.0 cm^3 e) 50.0 cm^3

49. Which of the following is NOT an SI base unit?

a) mass b) volume c) length d) time e) temperature

50. An iron "shotput" which is used in track meets has a mass of 16.00 lbs. (7260 g). If the density of iron is 7.87 g/cm^3, what is the radius of the iron ball? (The volume of a sphere is $4/3\ \pi\ r^3$.)

a. 3.02 cm b. 6.04 cm c. 7.31 cm d. 12.1 cm e. 24.2 cm

Chapter 1: Answers:

1. c	26. c
2. a	27. d
3. d	28. d
4. a	29. a
5. c	30. d
6. d	31. a
7. b	32. c
8. e	33. c
9. b	34. c
10. e	35. a
11. b	36. a
12. c	37. c
13. d	38. d
14. c	39. b
15. e	40. b
16. c	41. c
17. e	42. c
18. b	43. d
19. e	44. d
20. d	45. e
21. b	46. c
22. d	47. a
23. e	48. e
24. a	49. b
25. e	50. b

Chapter 2
Atoms and Elements

1. John Dalton's atomic postulates included all of the following EXCEPT

   a) All matter is composed of atoms.

   b) All atoms of a given element are identical, but atoms of each element are different.

   c) Compounds are formed by the combination of two or more kinds of elements in small whole number ratios.

   d) Chemical reactions are the result of electrons being transferred from one atom to another.

   e) Atoms are not created, destroyed, divided into parts, or converted into other kinds of atoms during a chemical reaction.

2. Which of the following sets of compounds illustrate the law of multiple proportions as set forth by John Dalton?

   a) CO_2, H_2O, N_2H_4, C_2H_4 b) NO, NO_2, N_2O, N_2O_5 c) CO_2, SO_2, SiO_2, TiO_2

   d) NO_2, NH_3, H_2NOH, H_2O e) C_2H_6, H_6C_2, CH_3CH_3, $(CH_3)_2$

3. If 4.00 g of element A react completely with 8.00 g of element B, which of the following statements must be is true?

   a) The reactions requires 4 moles of A to react with 8 moles of B.

   b) The maximum mass of product that can be formed is 12.00 g.

   c) The maximum amount of product that can be formed is 12 moles.

   d) The atomic weight of B is double the atomic weight of A.

   e) The density of B is greater than the density of A.

4. About 1910 Rutherford and colleagues performed experiments by targeting a stream of alpha particles at a piece of gold foil and recording the deflection of the particles on a sensitive screen. Which of the following statement(s) were conclusion(s) that were drawn from those experiments?

   1. Most of the volume of the atom is empty space.
   2. The nucleus of an atom is extremely dense.
   3. Electrons are negatively charged.

   a) 1 only b) 2 only c) 1 and 2 only d) 2 and 3 only e) 1, 2, and 3

5. In the Millikan oil drop experiment, the charge on oil droplets was observed by their behavior between a positively charged plate and a negatively charged plate. The fundamental charge on an electron was determined as -1.60×10^{-19} coulombs by observing that

a) the charge on all the droplets was a multiple of 1.60×10^{-19} coulombs.

b) the charge on all the droplets was -1.60×10^{-19} coulombs.

c) the charge on all the droplets was $+1.60 \times 10^{-19}$ coulombs.

d) the charge on all the droplets was $+1.60 \times 10^{19}$ coulombs.

e) the charge on all the droplets was -1.60×10^{19} coulombs

6. The number of isotopes in a sample of a pure element can best be determined experimentally with a(n)

a) electroscope. b) mass spectrometer. c) electron microscope. d) cathode ray tube.

e) scanning tunneling microscope.

7. The masses 9.1094×10^{-28} g, 1.6726×10^{-24} g, and 1.6749×10^{-24} g represent what particles respectively?

a) proton, neutron, electron b) electron, proton, neutron c) electron, neutron, proton

d) neutron, proton, electron e) proton, electron, neutron

8. Chlorine has two stable isotopes with masses of 34.97 amu and 36.97 amu. What is the relative abundance of the two isotopes?

a) 50.00% ^{35}Cl and 50.00% ^{37}Cl

b) 35.45% ^{35}Cl and 64.55% ^{37}Cl

c) 64.55% ^{35}Cl and 35.45% ^{37}Cl

d) 24.23% ^{35}Cl and 75.77% ^{37}Cl

e) 75.77% ^{35}Cl and 24.23% ^{37}Cl

9. Boron has two stable isotopes with masses of 10.01 amu and 11.01 amu. What is the relative abundance of the two isotopes?

a) 50.00% ^{10}B and 50.00% ^{11}B

b) 19.91% ^{10}B and 80.09% ^{11}B

c) 80.09% ^{10}B and 19.91% ^{11}B

d) 10.81% ^{10}B and 89.19% ^{11}B

e) 89.19% ^{10}B and 10.81% ^{11}B

10. Which of the following atoms contain 24 neutrons?

a) $^{52}_{24}Cr$ b) $^{24}_{12}Mg$ c) $^{112}_{48}Cd$ d) $^{48}_{24}Cr$ e) $^{42}_{20}Ca$

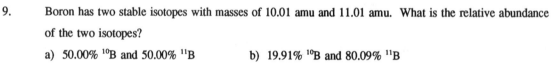

11. Copper has two stable isotopes, ^{63}Cu and ^{65}Cu. How many protons, electrons, and neutrons does an atom of ^{65}Cu contain?

	Protons	Electrons	Neutrons
a)	29	36	29
b)	36	29	29
c)	29	29	36
d)	36	29	29
e)	65	65	29

12. How many neutrons are in $^{79}_{34}Se$?

a) 34 b) 45 c) 79 d) 113 e) 117

13. Identify the set which contains isotopes of the same element.

a) $^{17}_{8}A$ and $^{17}_{9}A$ b) $^{12}_{7}A$ and $^{7}_{12}A$ c) $^{11}_{7}A$ and $^{12}_{7}A$ d) $^{11}_{5}A$ and $^{11}_{6}A^{+}$
e) $^{11}_{24}A$ and $^{24}_{11}A$

14. What is the mass number of an atom of iodine with 76 neutrons?

a) 76 b) 53 c) 106 d) 129 e) 258

15. Which of the following is **NOT** an element of the fourth period in the periodic table?

a) Co b) V c) Mg d) Ca e) Kr

16. A transition metal, a halogen, and a metalloid in that order are

a) Ni, N, Sn b) Sc, Si, Sb c) Cr, Cl, As d) As, Cl, Se e) Ca, Cl, Se

17. An alkaline earth, a chalcogen, and a noble gas in that order are

a) Na, S, Ne b) Ca, O, Ne c) K, Cl, Kr d) Sr, S, Se e) Mg, Br, Kr

18. Three elements in the lanthanide series are

a) Ce, U, Rn b) Ce, Nd, Sm c) Ce, Ta, Nb d) Cs, Ba, Ce e) Fr, Ra, Ce

19. Which of the following elements is a non-metal?

a) Ca b) Cr c) Co d) Cu e) Cl

20. A 1.00 g sample of sodium contains _____ moles.

a) 2.16×10^{23} b) 2.62×10^{22} c) 1.25×10^{25} d) 22.99 e) 0.0435

21. To obtain 1.50 moles of iron you must weigh _____ grams of iron.

a) 5.04×10^{25} b) 2.24×10^{25} c) 4.46×10^{-26} d) 37.2 e) 83.8

22. A sample of 1.00 g of lead contains _____ atoms.

a) 4.83×10^{-3} b) 1.25×10^{26} c) 6.02×10^{23} d) 207.2 e) 2.91×10^{21}

23. You have 2.00 g of each of the following: Se, Si, Sn, S, and Sc. Which contains the largest number of atoms?

a) Se b) Si c) Sn d) S e) Sc

24. You have 0.125 mole of each of the following elements: K, Al, C, Ca, and Cu. Which sample has the largest mass?

a) K b) Al c) C d) Ca e) Cu

25. Consider a gold coin which is 90.0% gold and 10.0% copper. If the coin has a mass of 2.00 ounces (56.7 g), how many moles of gold are in the coin?

a) 0.0892 mol b) 0.259 mol c) 0.288 mol d) 0.892 mol e) 25.9 mol

26. To obtain 1.20×10^{24} atoms of nickel, you would weigh

a) 2.00 g. b) 29.4 g. c) 117 g. d) 7.04×10^{25} g. e) 1.42×10^{-26} g.

27. To obtain 5.66×10^{21} atoms of nickel, you would weigh

a) 0.552 g. b) 1.81 g. c) 106 g. d) 9.64×10^{19} g. e) 1.04×10^{-20} g.

28. What is an expression for calculating the average mass of ***one*** atom of argon?

a) 39.9 g/atom b) 1 g/6.02×10^{23} atoms c) 39.9 g/6.02×10^{23} atoms

d) 6.02×10^{23} atoms/39.9 g e) (39.9 g/1 mol) (6.02×10^{23} atoms/1 mol)

29. What is an expression for calculating the average mass of ***one*** atom of calcium?

a) (40.08 g/1 mol)(6.022×10^{23} atoms/1 mol)

b) (1 mol/40.08 g)(6.022×10^{23} atoms/1 mol)

c) (40.08 g/1 mol)(1 mol/6.022×10^{23} atoms)

d) (1 mol/40.08 g)(1 mol/6.022×10^{23} atoms)

e) (40.08 g/1 atom)(6.022×10^{23} atoms/1 gram)

30. You have 1.36×10^{-4} g of the radioactive element americium, Am. How many moles of americium do you have?

a) 0.0331 mol b) 5.60×10^{-7} mol c) 1.79×10^{6} mol d. 4.11×10^{-3} mol e. 4.11×10^{-5} mol

31. An average sample of coal contains 3.0% sulfur by mass. How many moles of sulfur are there in 1.00 metric ton of coal? (1 metric ton = 1000. kg)

a) 940 mol b) 94 mol c) 31 mol d) 1.1 mol e. 0.94 mol

32. What is the average mass of one atom of copper?

a) 9.48×10^{-21} g b) 1.06×10^{-22} g c) 6.02×10^{-23} g d) 3.82×10^{25} g e) 1.66×10^{-24} g

33. Zinc has a density of 7.14 g/cm^3. If you have a piece of zinc that is 0.20 cm thick, 1.5 cm wide, and 3.0 cm long, how many moles of zinc are present?

a) 0.0098 mol b) 0.098 mol c) 0.21 mol d) 2.1 mol e) 21 mol

34. If 6.00 mL of argon which has a density of 1.78 g/L is used in an experiment, how many moles and how many atoms are used?

a) 2.67×10^{-7}mol, 1.61×10^{17} atoms

b) 1.34×10^{-2} mol, 6.21×10^{18} atoms

c) 1.07×10^{-5} mol, 6.43×10^{18} atoms

d) 1.18×10^{-11} mol, $1.97\ 10^{17}$ atoms

e) 4.45×10^{-8} mol, 2.68×10^{16} atoms

35. The density of copper is 8.96 g/cm^3 at room temperature. If a cube of copper weigh 25.0 grams, what is the length of its edge?

a) 0.635 cm b) 3.91 cm c) 6.79 cm d) 1.41 cm e) 2. 07 cm

36. The density of lead is 11.3 g/cm^3 at room temperature. How many atoms are in a cube of pure lead which has an edge of 2.00 cm ?

a) 4.11×10^{21} atoms b) 3.11×10^{20} atoms c) 1.38×10^{24} atoms

d) 2.63×10^{23} atoms e) 6.57×10^{22} atoms

37. The density of argon is 1.78 g/L at 0°C and 1 atm pressure. What volume of argon would contain 3.01×10^{25} atoms?

a) 447 L b) 22.4 L c) 1120 L d) 560 L e) 13.7 L

38. How many grams of magnesium contain the same number of atoms as 20.04 g of calcium?

a) 12.16 g b) 20.04 g c) 24.30 g d) 40.08 g e) 48.60 g

39. How many grams of magnesium contain the same number of atoms as 1.00 g of calcium?

a) 0.411 g b) 0.607 g c) 1.65 g d) 1.08×10^{-24} g e) 9.93×10^{-23} g

40. How many grams of iron contain the same number of atoms as 50.0 g of aluminum?

a) 1.24×10^{24} g b) 1.08×10^{-24} g c) 80.2 g d) 103 g e) 111 g

41. You have set up an experiment in which you will react mercury with sulfur. If you have 0.64 g of S and 0.64 g of Hg, which statement below best describes the situation?

a) There are equal numbers of moles of each element present.

b) There are more moles of S than moles of Hg present.

c) There are more moles of Hg than moles of S present.

d) There are equal numbers of atoms of each element present.

e) There are more atoms of Hg than atoms of S present.

42. A flask contains 40 g of neon and 40 g of argon. Which of the following statement(s) is (are) true?

1. There are more moles of Ar than moles of Ne present.
2. There are more atoms of Ne than atoms of Ar present.
3. There are equal numbers of each element present.

a) 1 only b) 2 only c) 3 only d. 1 and 2 only e. 1 and 3 only

43. The positive charge in the nucleus of an element determines the

a) atomic mass. b) mass number. c) atomic number. d) number of neutrons.

e) radioactivity.

44. The number of neutrons in 30 molecules of As_4 where As has the mass number of 75 is

a) 9000. b) 6720. c) 5040. d) 3960. e) 1760.

45. There are two stable isotopes of carbon. They differ with respect to

a) atomic mass. b) number of protons. c) radioactivity. d) atomic number.

e) electron configuration.

46. A transition metal, a halogen, and a metalloid in that order are

a) Ni, N, Sn. b) Sc, Si, Sb. c) As, Cl, Se. d) Bi, Br, C. e) Cr, Cl, As.

47. A lanthanide, a chalcogen, and a transition metal in that order are

a) Tb, Tl, Tc. b) Sm, S, Sc. c) Cm, Cl, Cs. d) U, O, Os. e) Am, As, Au.

48. Which of the following elements are in the same chemical family?

a) Rn, Ba, Sr, Be b) N, O, F, Ne c) Li, Be, Na, Mg d) Ge, As, Sb, Te e) Si, Sn, C, Pb

49. The chemical properties of sulfur would be most similar to

a) P. b) Cl. c) Ar. d) Se. e) As.

50. The chemical properties of germanium would be most similar to

a) P. b) Ga. c) Si. d) Sb. e) Ho.

Chapter 2: Answers:

1. d	26. c
2. b	27. a
3. b	28. c
4. c	29. c
5. a	30. b
6. b	31. a
7. b	32. b
8. e	33. b
9. b	34. a
10. d	35. d
11. c	36. d
12. b	37. c
13. c	38. a
14. d	39. b
15. c	40. d
16. c	41. b
17. b	42. b
18. b	43. c
19. e	44. c
20. e	45. a
21. e	46. e
22. e	47. b
23. b	48. e
24. e	49. d
25. b	50. c

Chapter 3
Molecules and Compounds

1. How many atoms are in 12 molecules of fructose, $C_6H_{12}O_6$?

a) 24 b) 288 c) 2160 d) 7.22×10^{24} e) 1.73×10^{26}

2. In 0.50 mole of methyl formate, $HCOOCH_3$, there are

a) 6.0×10^{23} molecules. b) 1.2×10^{24} molecules. c) 1.8×10^{24} atoms. d) 2.4×10^{24} atoms.
e) 4.8×10^{24} atoms.

3. All of the following describe 58.1 g of butane, C_4H_{10}, **EXCEPT**

a) one mole of butane.
b) The amount of butane that contains 10.1 g of hydrogen.
c) The amount of butane that contains $10 \times 6.02 \times 10^{23}$ hydrogen atoms.
d) The amount of butane that contains 48.0 g of carbon.
e) $58.1 \times 6.02 \times 10^{23}$ molecules of butane.

4. In 0.250 moles of ethylene glycol (antifreeze), $HOCH_2CH_2OH$, there are

a) 1.51×10^{23} atoms. b) 1.51×10^{24} molecules. c) 1.51×10^{24} atoms. d) 6.02×10^{24} atoms.
e) 3.01×10^{24} molecules.

5. How many moles of fluorine **molecules** are in 5.00 grams of elemental fluorine?

a) 0.132 mol b) 0.263 mol c) 3.80 mol d) 1.07×10^{23} mol e) 5.35×10^{23} mol

6. How many molecules are in 75.0 grams of nitrogen gas?

a) 6.45×10^{24} molecules b) 1.26×10^{27} molecules c) 8.06×10^{24} molecules
d) 1.61×10^{24} molecules e) 6.32×10^{26} molecules

7. How many moles are in 8.50 mL of liquid bromine if the density of bromine is 3.12 g/mL?

a) 0.332 mol b) 0.166 mol c) 0.0830 mol d) 0.0415 mol e) 0.0207 mol

8. A sample of oxygen gas weighs 32.0 grams. It contains 6.022×10^{23} _____.

a) protons b) neutrons c) electrons d) molecules e) atoms

9. A mole of chlorine **molecules** weighs _____ grams and contains _____ atoms.

a) 70.91 grams and 6.022×10^{23} atoms
b) 70.91 grams and 12.04×10^{23} atoms
c) 35.45 grams and 6.022×10^{23} atoms
d) 35.45 grams and 12.04×10^{23} atoms
e) 35.45 grams and 3.011×10^{23} atoms

10. How many moles of potassium carbonate are in a 15.0 g K_2CO_3 sample?

a) 6.53×10^{22} mol b) 2.07×10^{23} mol c) 0.151 mol d) 0.109 mol e) 9.20 mol

11. How many moles of SO_2 are in 1.07×10^{23} molecules of SO_2?

a) 11.4 mol b) 0.178 mol c) 1.67×10^{21} mol d) 2.77×10^{-3} mol e) 6.85×10^{24} mol

12. How many grams of CH_4 contain the same number of molecules as 2.50 grams O_2?

a) 1.25 g b) 0.0781 g c) 0.156 g d) 4.70×10^{22} g e) 4.88×10^{-3} g

13. How many grams of NO_2 contain the same number of molecules as 5.00 grams H_2O?

a) 12.8 g b) 2.77 g c) 5.00 g d) 0.391 g e) 1.96 g

14. Which of the following has the same number of atoms as 55.8 grams of iron?

a) 55.8 g H_2 b) 55.8 g He c) 32.0 g O_2 d) 16.0 g O_2 e) 32.0 g O_3

15. Which of the following contains the largest number of molecules: 6.00 g CH_4, 9.00 g H_2O, 15.0 g NO_2, 11.0 g C_2H_6, or 20.0 g C_2H_5OH?

a) CH_4 b) H_2O c) NO_2 d) C_2H_6 e) C_2H_5OH

16. Which of the following compounds will form a solution with water that is a good conductor of electricity?

a) CCl_4 b) CO_2 c) NaCl d) Cl_2 e) CH_3OH

17. Which of the following compounds will form a solution with water that is a good conductor of electricity?

a) CH_4 b) C_2H_6 c) $CaCl_2$ d) SiH_4 e) CH_3OH

18. Which of the following series represents only known stable metal ions?

a) Fe^{2+}, Fe^{3+}, K^{2+} b) Mg^{2+}, Ba^{3+}, Na^{+} c) Li^{2+}, Na^{+}, Al^{3+} d) Fe^{2+}, Sr^{2+}, Mg^{2+}
e) Li^{2+}, Ca^{2+}, Al^{3+}

19. Which of the following series represents only known stable nonmetal ions?

a) O^{2-}, Cl^{2-}, K^{-} b) O^{2-}, Cl^{2-}, N^{3-} c) S^{2-}, N^{2-}, Cl^{-} d) O_2^{-}, O^{2-}, P^{2-} e) S^{2-}, P^{3-}, F^{-}

20. The calcium ion, Ca^{2+}, has

a) two more electrons than the calcium atom

b) two less electrons than the calcium atom

c) two more protons than the calcium atom

d) two less protons than the calcium atom

e) two more electrons and two more protons than the calcium atom

21. The fluoride ion has _____ electrons.

a) 8 b) 9 c) 10 d) 19 e) 20

22. The following species F^-, Ne, Na^+, and Mg^{2+} all have the same number of

a) protons. b) neutrons. c) electrons. d) nucleons. e) charges.

23. For an atom from Group 7A of the Periodic Table, the most common monatomic ion of this atom will have a charge of

a) +7 b) -7 c) +8 d) -1 e) -2

24. Which of the following pairs have the same number of electrons?

a) Fe^{2+} and Fe^{3+} b) Ca^{2+} and K^+ c) K^+ and Na^+ d) O_2^- and O^{2-} e) H^+ and H^-

25. Which of the following pairs have the same number of electrons?

a) Cu^{2+} and Cu^+ b) Ca^{2+} and Mg^{2+} c) Cl^- and Br^- d) F^- and O^{2-} e) H^+ and H^-

26. Which formula represents the binary compound formed by magnesium and nitrogen?

a) MgN b) Mg_2N c) MgN_3 d) Mg_3N_2 e) Mg_2N_3

27. Which formula represents the binary compound formed by sodium and tellurium?

a) Na_2Te b) NaTe c) Na_2Te d) Na_3Te_2 e) Na_3Te

28. Which group of compounds are ALL ionic?

a) H_2O, NaCl, CS_2 b) NaCl, CH_4, $CaCl_2$ c) $CaCl_2$, $FeCl_3$, NaCl d) H_2O, $FeCl_3$, CO_2

e) $CaCl_2$, $FeCl_3$, CO_2

29. Give the ions present and their relative numbers in potassium sulfate.

a) 1 K^+ and 1 SO_4^- b) 2 K^+ and 1 SO_3^{2-} c) 1 K^+ and 2 SO_4^{2-} d) 2 K^+ and 1 SO_4^{2-} e) 3 K^+ and 1 SO_4^{3-}

30. Give the ions present and their relative numbers in barium nitrate.

a) 1 Ba^+ and 1 NO_3^- b) 2 Ba^+ and 1 NO_3^{2-} c) 1 Ba^{2+} and 2 NO_3^- d) 1 Ba^{2+} and 1 NO_2^{2-}

e) 1 Ba^{2+} and 2 NO_2^{2-}

31. What is the name of $FeSO_4$?

a) iron(III) sulfite b) iron(II) sulfite c) iron(II) sulfate d) iron(II) sulfide e) iron(III) sulfide

32. What is the correct name of $KClO_4$?

a) potassium chlorate b) potassium perchlorite c) potassium perchlorate d) potassium hypochlorate

e) potassium hypochlorite

33. What is the formula for sodium bicarbonate?

a) $Na(CO_3)_2$ b) $NaHCO_3$ c) Na_2CO_3 d) $NaCO_2$ e) $NaHCO_2$

34. The formula of the compound ammonium phosphate is

a) $NH_4(PO_4)_3$ b) $NH_4(PO_4)_2$ c) $(NH_4)_2 PO_4$ d) $(NH_4)_3PO_4$ e) $(NH_4)_3(PO_4)_2$

35. The formula of barium molybdate is $BaMoO_4$. Therefore, the formula of sodium molybdate is

a) Na_4MoO. b) $NaMoO$. c) Na_2MoO_3 d) Na_2MoO_4 e) Na_4MoO_4

36. Sodium oxalate has the chemical formula, $Na_2C_2O_4$. Based on this information, the formula of calcium oxalate is

a) Ca_2CO_3. b) CaC_2O_4. c) $Ca_2C_2O_4$ d) $CaHCO_3$ e) $CaCO_3$

37. What is the weight percent of carbon in oxalic acid, $H_2C_2O_4$?

a) 2.24% b) 13.34% c) 26.68% d) 34.52% e) 42.18%

38. Which of the following compounds is 36.4% oxygen by mass?

a) N_2O b) NO c) N_2O_3 d) N_2O_4 e) N_2O5

39. The molar mass of barium nitrate is

a) 199.21 g/mol. b) 229.32 g/mol. c) 261.35 g/mol. d) 336.61 g/mol. e) 398.62 g/mol.

40. The molar mass of sodium chlorate is

a) 58.44 g/mol. b) 81.44 g/mol. c) 93.90 g/mol. d) 106.44 g/mol. e) 129.44 g/mol.

41. The molar mass of barium hydroxide octahydrate is

a) 189.34 g/mol. b) 281.38 g/mol. c) 298.38 g/mol. d) 299.48 g/mol. e) 315.48 g/mol.

42. The molar mass of a compound with an empirical formula of BH_3 is 27.67 g/mol. What is the molecular formula?

a) B_2H_6 b) B_2H_3 c) B_3H_6 d) B_2H_4 e) B_2H_5

43. All of the compounds listed below have a molar mass of 240 amu. Which of the listed compounds could have an empirical mass of 80 amu?

a) $C_{14}H_8O_4$ b) $C_{10}H_8OS_3$ c) $C_{15}H_{12}O_3$ d) $C_6H_{12}N_2S_4$ e) $C_6H_3Cl_3N_2O_2$

44. Maleic acid, which is used to manufacture artificial resins, has the empirical formula CHO. Its molar mass is 116.1 g/mol. What is it molecular formula?

a) CHO b) $C_2H_2O_2$ c) $C_3H_3O_3$ d) $C_4H_4O_4$ e) $C_6H_6O_6$

45. Nitrogen and oxygen form an extensive series of at least seven oxides with the general formula N_xO_y. One of them is a blue solid that comes apart, reversibly, in the gas phase. It contains 36.84% N. What is the empirical formula of the oxide?

a) N_2O b) NO c) N_2O_3 d) NO_2 e. N_2O_5

46. Boron forms an extensive series of hydrides. If a hydride is isolated that is 81.1% B, a possible molecular formula of the compound is

a) B_2H_6 b) B_3H_9 c) B_4H_{10} d) B_5H_9 e) B_5H_{11}

47. A compound contains 15.94% boron with the remainder being fluorine. What is the empirical formula of the compound?

a) BF b) BF_2 c) B_2F_3 d) BF_3 e. B_2F_5

48. Calculate the empirical formula of a compound which is 51.40% carbon, 8.63% hydrogen, and 39.97% nitrogen.

a) $C_5H_8N_2$ b) $C_4H_9N_3$ c) C_2H_3N d) $C_4H_6N_2$ e) $C_3H_6N_2$

49. Transition metals can combine with carbon monoxide (CO) to form compounds such as $Fe(CO)_5$ and $Co_2(CO)_8$. Assume that you combine 0.125 g of nickel with CO and isolate 0.364 g of $Ni(CO)_x$. What is the value of x?

a) 1 b) 2 c) 3 d) 4 e) 6

50. A 3.26 g sample of an organic compound was found to contain 2.42 g carbon, 0.282 g hydrogen, and 0.563 g nitrogen. Calculate the empirical formula of the compound.

a) $C_5H_4N_3$ b) $C_{10}H_2N_3$ c) $C_6H_7N_4$ d) C_2H_3N e) C_5H_7N

Chapter 3: Answers:

1. b	26. d
2. d	27. c
3. e	28. c
4. c	29. d
5. a	30. c
6. d	31. c
7. b	32. c
8. d	33. b
9. b	34. d
10. d	35. d
11. b	36. b
12. a	37. c
13. a	38. a
14. d	39. c
15. b	40. d
16. c	41. e
17. c	42. a
18. d	43. c
19. e	44. d
20. b	45. c
21. c	46. c
22. c	47. d
23. d	48. e
24. b	49. d
25. d	50. e

Chapter 4
Chemical Equations and Stoichiometry

1. In a balanced chemical equation, what is balanced?

   a) atoms b) moles c) molecules d) atoms and molecules e) moles and atoms

2. Phosphorus trichloride may be prepared by the reaction of phosphorus with chlorine gas according to the equation below.

   $$\_\_ P_4 + \_\_ Cl_2 \rightarrow \_\_ PCl_3$$

   When the equation above is properly balanced with the smallest whole numbers, the respective coefficients are:

   a) 2, 6, 8 b) 1, 3, 4 c) 2, 3, 2 d) 1, 6, 4 e) 3, 9, 3

3. Carbon disulfide, CS_2, can be made from the reaction of graphite and SO_2.

   $$\_\_ C + \_\_ SO_2 \rightarrow \_\_ CS_2 + \_\_ CO$$

   When the equation above is properly balanced with the smallest whole numbers, the respective coefficients are:

   a) 2, 1, 1, 2 b) 5, 2, 3, 2 c) 4, 2, 1, 4 d) 3, 3, 1, 2 e) 5, 2, 1, 4

4. When glucose undergoes complete combustion, the products are carbon dioxide and water.

   $$\_\_ C_6H_{12}O_6 + \_\_ O_2 \rightarrow \_\_ CO_2 + \_\_ H_2O$$

   When the equation above is properly balanced with the smallest whole numbers, the respective coefficients are:

   a) 1, 9, 6, 6 b) 1, 6, 6, 6 c) 2, 12, 6, 12 d) 2, 12, 12, 12 e) 1, 9, 12, 12

5. Ammonia will react with fluorine at a high temperature in the presence of copper to produce some dinitrogen tetrafluoride and hydrogen fluoride.

   $$\_\_ NH_3(g) + \_\_ F_2(g) \rightarrow \_\_ N_2F_4(g) + \_\_ HF(g)$$

   When the equation above is properly balanced with the smallest whole numbers, the respective coefficients are:

   a) 2, 1, 1, 6 b) 2, 3, 1, 6 c) 2, 5, 1, 6 d) 2, 10, 1, 6 e) 2, 6, 1, 6

6. The balanced equation for the combustion of butane, C_4H_{10}, is

a) $C_4H_{10} + 13O_2 \rightarrow 4CO_2 + 5H_2O$

b) $2C_4H_{10} + 13O_2 \rightarrow 8CO_2 + 10\ H_2O$

c) $2C_4H_{10} + 23O_2 \rightarrow 8CO_2 + 5H_2O$

d) $C_4H_{10} + 16O_2 \rightarrow 4CO_2 + 10H_2O$

e) $C_4H_{10} + 23O_2 \rightarrow 16CO_2 + 5H_2O$

7. The balanced equation for the complete combustion of cyclohexane, C_6H_{12}, is

a) $C_6H_{12} + 18O_2 \rightarrow 6CO_2 + 6H_2O$

b) $C_6H_{12} + 9O_2 \rightarrow 6CO_2 + 6H_2O$

c) $C_6H_{12} + 6O_2 \rightarrow 6CO_2 + 6H_2$

d) $C_6H_{12} + 6O_2 \rightarrow 6CO_2 + 6H_2O$

e) $C_6H_{12} + 12O_2 \rightarrow 6CO_2 + 6H_2$

8. Ammonia can be made by reaching calcium cyanamide with water.

$$\_\_\ CaCN_2(s) + \_\_\ H_2O(\ell) \rightarrow \_\_\ CaCO_3(s) + \_\_\ NH_3(g)$$

When the equation above is properly balanced with the smallest whole numbers, the respective coefficients are

a) 2, 6, 2, 4 b) 4, 2, 6, 2 c) 1, 3, 1, 2 d) 1, 1, 1, 1 e) 1, 4, 2, 2

9. The balanced equation for the combustion of ferrocene, $C_{10}H_{10}Fe$, in oxygen to give iron(III) oxide, carbon dioxide, and water is

a) $C_{10}H_{10}Fe(s) + O_2(g) \rightarrow 2Fe_2O_3(s) + 10CO_2(g) + 5H_2O(g)$

b) $C_{10}H_{10}Fe(s) + 13O_2(g) \rightarrow FeO(s) + 10CO_2(g) + 5H_2O(g)$

c) $2C_{10}H_{10}Fe(s) + 26O_2(g) \rightarrow Fe_2O_3(s) + 20CO_2(g) + 10H_2O(g)$

d) $4C_{10}H_{10}Fe(s) + 53O_2(g) \rightarrow 2Fe_2O_3(s) + 40CO_2(g) + 20H_2O(g)$

e) $2C_{10}H_{10}Fe(s) + O_2(g) \rightarrow Fe_2O_3(s) + CO_2(g) + 10H_2O(g)$

10. The balanced equation for the combustion of isopropyl alcohol in air is

a) $C_3H_7OH(\ell) + 5O_2(g) \rightarrow 3CO_2(g) + 4H_2O(g)$

b) $2C_3H_7OH(\ell) + 9O_2(g) \rightarrow 6CO_2(g) + 8H_2O(g)$

c) $C_3H_7OH(\ell) + 3O_2(g) \rightarrow 3CO(g) + 4H_2O(g)$

d) $C_3H_7OH(\ell) + 9O(g) \rightarrow 3CO_2(g) + 4H_2O(g)$

e) $C_3H_7OH(\ell) + 10\ O_2(g) \rightarrow 6CO_2(g) + 8H_2O(g)$

11. The balanced equation for the reaction of lithium with oxygen is

a) $4Li(s) + O_2(g) \rightarrow 2Li_2O(s)$ b) $2Li(s) + O_2(g) \rightarrow 2LiO(s)$ c) $2Li(s) + O(g) \rightarrow Li_2O(s)$

d) $4Li(s) + 2O_2(g) \rightarrow 2Li_2O(s)$ e) $Li(s) + O_2(g) \rightarrow LiO_2(g)$

12. The balanced equation for the reaction of chromium with oxygen to give chromium(III) oxide is

a) $Cr(s) + O_2(g) \rightarrow CrO_2(s)$ b) $3Cr(s) + O_2(g) \rightarrow Cr_3O_2(s)$ c) $Cr(s) + O_2(g) \rightarrow Cr_2O_3(s)$

d) $4Cr(s) + 3O_2(g) \rightarrow 2Cr_2O_3(s)$ e) $2Cr(s) + O_2(g) \rightarrow 2CrO(s)$

13. The balanced equation for the reaction of sulfur with oxygen is

a) $S_8(s) + 16O(g) \rightarrow 8SO_2(g)$ b) $S_8(s) + 8O_2(g) \rightarrow 8SO_2(g)$ c) $S_8(s) + 32O_2(g) \rightarrow 8SO_4(g)$

d) $2S_8(s) + 8O_2(g) \rightarrow 8S_2O_2(g)$ e) $S_8(s) + 4O_2(g) \rightarrow S_8O_8(s)$

14. Nitrogen oxide is oxidized in air to give brown nitrogen dioxide.

$$2NO(g) + O_2(g) \rightarrow 2NO_2(g)$$

If you have 2.2 moles of NO,

a) you need 2.2 moles of O_2 for complete reaction and produce 2.2 moles of NO_2.

b) you need 1.1 moles of O_2 for complete reaction and produce 2.2 moles of NO_2.

c) you need 1.1 moles of O_2 for complete reaction and produce 3.3 moles of NO_2.

d) you need 1.0 moles of O_2 for complete reaction and produce 2.0 moles of NO_2.

e) you need 2.2 moles of O_2 for complete reaction and produce 4.4 moles of NO_2.

15. Aluminum reacts with oxygen to give aluminum(III) oxide.

$$4Al(s) + 3O_2(g) \rightarrow 2Al_2O_3(s)$$

If you have 6.0 moles of Al,

a) you need 3.0 moles of O_2 for complete reaction and produce 2.0 moles of Al_2O_3.

b) you need 18.0 moles of O_2 for complete reaction and produce 12.0 moles of Al_2O_3.

c) you need 3.0 moles of O_2 for complete reaction and produce 2.0 moles of Al_2O_3.

d) you need 4.5 moles of O_2 for complete reaction and produce 3.0 moles of Al_2O_3.

e) you need 4.5 moles of O_2 for complete reaction and produce 2.0 moles of Al_2O_3.

16. Many metals react with halogens to give metal halides. For example,

$$Fe(s) + Cl_2(g) \rightarrow FeCl_2(s)$$

If you begin with 10.0 g of iron,

a) you will need 10.0 g of Cl_2 for complete reaction and will produce 22.7 g of $FeCl_2$.

b) you will need 12.7 g of Cl_2 for complete reaction and will produce 10.0 g of $FeCl_2$.

c) you will need 12.7 g of Cl_2 for complete reaction and will produce 22.7 g of $FeCl_2$.

d) you will need 10.0 g of Cl_2 for complete reaction and will produce 10.0 g of $FeCl_2$.

e) you will need 10.0 g of Cl_2 for complete reaction and will produce 20.0 g of $FeCl_2$.

17. The very stable compound SF_6 is made by burning sulfur in an atmosphere of fluorine.

$$S_8(s) + 24F_2(g) \rightarrow 8SF_6(g)$$

If you need 2.50 moles of SF_6, you will need to use

a) 0.313 moles of S_8 and 7.50 moles of F_2.

b) 1.00 moles of S_8 and 24.0 moles of F_2.

c) 0.125 moles of S_8 and 3.00 moles of F_2.

d) 8.00 moles of S_8 and 24.0 moles of F_2.

e) 0.125 moles of S_8 and 7.50 moles of F_2.

18. How many grams of carbon are needed to react completely with 75.2 grams of SiO_2 according to the following equation?

$$SiO_2(s) + 3C(s) \rightarrow SiC(s) + 2CO(g)$$

a) 15.0 g b) 20.5 g c) 32.8 g d) 45.1 g e) 61.5 g

19. How many grams of bromine are needed to react completely with 85.0 grams of NH_3 to produce ammonium bromide and nitrogen according to the equation below?

$$3Br_2(\ell) + 8NH_3(g) \rightarrow 6NH_4Br(s) + N_2(g)$$

a) 247 g b) 337 g c) 447 g d) 587 g e) 897 g

20. If 150.0 g SiO_2 and 60.0 grams C react according to the equation below, what is the maximum number of moles of CO that can be produced?

$$SiO_2(s) + 3C(s) \rightarrow SiC(s) + 2CO(g)$$

a) 3.33 mol b) 3.67 mol c) 5.00 mol d) 6.5 mol e) 7.50 mol

21. If 0.195 mol of bromine and 1.18 mol of ammonia are combined and react according to the equation below, what is the maximum number of grams of N_2 that can be produced?

$$3Br_2(\ell) + 8NH_3(g) \rightarrow 6NH_4Br(s) + N_2(g)$$

a) 1.62 g b) 1.82 g c) 2.42 g d) 4.11 g e) 4.76 g

22. The compound P_4S_3 is used in matches and its reaction with oxygen is

$$P_4S_3(s) + 8O_2(g) \rightarrow P_4O_{10}(s) + 3SO_2(g)$$

How many grams of O_2 are needed to react with 0.450 grams of P_4S_3? (Molar mass of P_4S_3 is 220.1 g/mol.)

a) 0.0654 g b) 0.261 g c) 0.523 g d) 3.60 g e) 1.83 g

23. Given the unbalanced equation: $Fe + O_2 \rightarrow Fe_2O_3$

The number of moles of oxygen gas which will react with 1.6 mol of iron to produce iron(III) oxide is

a) 1.2 mol b) 1.6 mol c) 2.2 mol d) 2.4 mol e) 3.2 mol

24. Water is produced by burning hydrogen.

$$2H_2 + O_2 \rightarrow 2H_2O + \text{heat}$$

If 4.0 g of hydrogen gas and 8.0 g of oxygen gas are available as reactants, what is the limiting reagent, if any?

a) H_2 b) O_2 c) heat d) H_2O e) there is no limiting reagent

25. Given the balanced equation: $A + 3B \rightarrow 2C$.

The molar mass of C is 40.0 grams. If one **mole** of A produces 20.0 grams of C, what is the percent yield of the reaction?

a) 100% b) 50% c) 25% d) 10% e) 1.0%

26. How many grams of carbon dioxide are produced by the complete combustion of 5.00 grams of acetylene?

$$2C_2H_2(g) + 5O_2(g) \rightarrow 4CO_2(g) + 2H_2O(g)$$

(The molar masses are C_2H_2 = 26.04 g/mol and CO_2 = 44.01 g/mol.)

a) 8.45 g b) 16.9 g c) 20.0 g d) 33.8 g e) 42.6 g

27. How many grams of carbon dioxide are produced by the complete combustion of 5.00 grams of pentane?

$$C_5H_{12}(\ell) + 8O_2(g) \rightarrow 5CO_2(g) + 6H_2O(\ell$$

(The molar masses are C_5H_{12} = 72.15 g/mol and CO_2 = 44.01 g/mol.)

a) 0.346 g b) 3.04 g c) 4.15 g d) 12.7 g e) 15.2 g

28. How many moles of magnesium (if any) remain when 5.00 grams of magnesium is burned in 2.50 grams of pure oxygen?

$$2Mg(s) + O_2(g) \rightarrow 2MgO(s)$$

a) 0.206 mol b) 0.156 mol c) 0.0781 mol d) 0.0498 mol e) zero, it is totally consumed.

29. When 10.0 grams of mercury(II) oxide was decomposed, a student obtained 5.00 grams of mercury. What was the percent yield?

$$2HgO(s) \rightarrow 2Hg(\ell) + O_2(g)$$

a) 10.0% b) 33.3% c) 46.0% d) 50.0% e) 54.0%

30. Hydrazine, N_2H_4, is an important industrial reagent. It is synthesized by the Raschig process.

$$2NaOH(aq) + Cl_2(g) + 2NH_3(aq) \rightarrow N_2H_4(\ell) + 2NaCl(aq) + 2H_2O(\ell)$$

If you combine 100. g each of NaOH, Cl_2, and NH_3 and the reaction is complete, which reactant(s) will be left over?

1. NaOH
2. Cl_2
3. NH_3

a) 1 only b) 2 only c) 3 only d) 1 and 2 only e) 2 and 3 only

31. Phosphoric acid is made from phosphate rock, one form of which is apatite, $Ca_5(PO_4)_3F$ (molar mass = 504.3 g/mol).

$$Ca_5(PO_4)_3F(s) + 5H_2SO_4(aq) \rightarrow 5CaSO_4(s) + 5CaSO_4(s) + 3H_3PO_4(aq) + HF(aq)$$

If you use 100. g of apatite and 500. g of sulfuric acid (molar mass = 98.07 g/mol), what is the maximum possible yield of phosphoric acid (97.99 g/mol)?

a) 19.4 g b) 49.6 g c) 58.3 g d) 300. g e) 600. g

32. Ammonia gas can be prepared by the reaction of a basic oxide like calcium oxide with ammonium chloride, an acidic salt.

$$CaO(s) + 2NH_4Cl(s) \rightarrow 2NH_3(g) + H_2O(g) + CaCl_2(s)$$

If you isolate exactly 100. g of NH_3, but should have isolated 136 g in theory, what is the percentage yield of ammonia?

a) 36.8% b) 71.2% c) 73.5% d) 81.2% e) 90.0%

Questions 33-35 refer to the combustion of the hydrocarbon, decane.

33. Complete combustion of decane, $C_{10}H_{22}$, yields carbon dioxide and water. When the equation is balanced, using smallest whole number coefficients, the sum of the coefficients for the reactants is

a) 23. b) 29. c) 33. d) 42 e) 50

34. What is the maximum number of moles of water that can be produced from 56.92 g of $C_{10}H_{22}$?

a) 2.2 b) 2.8 c) 3.6 d) 4.0 e) 4.4

35. How many grams of oxygen are needed to burn the 56.92 g of $C_{10}H_{22}$?

a) 100 g b) 150 g c) 175 g d) 200 g e) 400 g

36. When a 75.5 gram sample of sulfur-containing coal was burned, 4.30 grams of SO_2 was produced. Assuming the coal to be a mixture of carbon and sulfur, what is the % sulfur in the coal?

a) 1.00% b) 1.43% c) 2.85% d) 5.70% e) 7.92%

37. The reaction below was known to proceed at 85.3% yield of SO_2.

$$P_4S_5(s) + 5O_2(g) \rightarrow P_4O_{10}(s) + 5SO_2(g)$$

(Molar mass P_4S_5 = 284.2 g/mol)

How many grams of P_4S_5 must be burned to produce 50.0 grams of SO_2?

a) 10.4 g b) 15.1 g c) 44.3 g d) 46.8 g e) 52.0 g

38. When 5.00 grams of Na_2CO_3 was treated with excess HCl, 2.00 grams of CO_2 was obtained. What was the % yield?

a) 40% b) 60% c) 83.4% d) 92.3% e) 96.3%

39. When 8.00 g of hydrogen reacts with 32.0 g of oxygen, the final mixture will contain

a) H_2, H_2O, O_2. b) H_2, H_2O. c) O, H_2O. d. H_2, O_2. e) pure H_2O.

40. A 10.0 g sample of an oxide of copper, when heated in a stream of hydrogen, forms 1.26 g of water. The % oxygen (by weight) in the compound is

a) 11.2% b) 12.6% c) 20.1% d) 79.9% e) 88.8%

41. Sulfur trioxide is made from the oxidation of sulfur dioxide and is represented by the equation

$$2SO_2(g) + O_2(g) \rightarrow 2SO_3(g)$$

A 16 g sample of SO_2 gives 18 g of SO_3. The **percent yield** of SO_3 is

a) 60% b) 75% c) 80% d) 90% e) 100%

42. Nitric oxide is made from the oxidation of ammonia and is represented by the equation

$$4NH_3(g) + 5O_2(g) \rightarrow 4NO(g) + 6H_2O(g)$$

An 8.5 g sample of NH_3 produced 12.0 g of NO. The **percent yield** of NO is (molar masses NH_3 = 17.0 g/mol and NO = 30.0 g/mol.)

a) 40% b) 60% c) 70% d) 80% e) 90%

43. Titanium(IV) oxide may be treated with chlorine and carbon to form titanium(IV) chloride with the release of carbon monoxide according to the following equation:

$$TiO_2(s) + 2Cl_2(g) + 2C(s) \rightarrow TiCl_4(g) + 2CO(g)$$

(Molar mass TiO_2 = 79.9 g/mol)

Suppose 50.0 grams of $TiO_2(s)$ is reacted with excess $Cl_2(g)$ and C(s) and 22.0 grams of CO is isolated. Calculate the percentage yield of CO.

a) 17.5% b) 35.1% c) 62.7% d) 83.5% e) 85.9%

44. 1.056 g of metal carbonate, containing an unknown metal M, were heated to give the metal oxide and 0.376 g CO_2.

$$MCO_3(s) + \text{heat} \rightarrow MO(s) + CO_2(g)$$

What is the identity of the metal M?

a) M = Ni b) M = Cu c) M = Zn d) M = Ba e) M = Mg

45. What is the empirical formula of a compound which is 64.80% carbon, 6.35% hydrogen, and 28.83% sulfur?

a) C_5H_6S b) $C_6H_7S_2$ c) C_5H_7S d) C_6H_7S e) $C_5H_5S_2$

46. Styrene, the building block of polystyrene, is a hydrocarbon, a compound consisting only of C and H. A given sample is burned completely and it produces 1.481 g of CO_2 and 0.303 g of H_2O. Determine the empirical formula of the compound.

a) CH b) CH_2 c) C_2H_3 d) C_2H_5 e) CH_3

47. Mesitylene is a liquid hydrocarbon. A given sample is burned completely and it produces 17.0 g of CO_2 and 4.64 g of H_2O. Determine the empirical formula of the compound.

a) CH b) CH_2 c) C_2H_3 d) CH_3 e) C_3H_4

48. A 27.0 g sample of an unknown carbon-hydrogen compound was burned in excess oxygen to form 88.0 g of CO_2 and 27.0 g H_2O. What is a possible molecular formula of the hydrocarbon?

a) CH_4 b) C_2H_6 c) C_4H_6 d) C_4H_8 e) C_4H_{10}

49. When $CaSO_4 \cdot XH_2O$ is heated, all of the water is driven off. If 34.0 g of $CaSO_4$ (molar mass = 136 g/mol) is formed from 43.0 g of $CaSO_4 \cdot XH_2O$, what is the value of X?

a) 1 b) 2 c) 3 d) 4 e) 5

50. A compound contains by weight 41.4% carbon, 3.47% H, and 55.1% oxygen. A 0.050-mole sample of this compound weighs 8.71 g. The molecular formula of the compound is:

a) CHO b) C_3H_3O c) $C_3H_3O_3$ d) $C_4H_4O_4$ e) $C_6H_6O_6$

Chapter 4: Answers:

1. a	26. b
2. d	27. e
3. e	28. d
4. b	29. e
5. c	30. e
6. b	31. c
7. b	32. c
8. c	33. c
9. d	34. e
10. b	35. d
11. a	36. c
12. d	37. e
13. b	38. e
14. b	39. b
15. d	40. a
16. c	41. d
17. a	42. d
18. d	43. c
19. b	44. b
20. a	45. d
21. b	46. a
22. c	47. e
23. a	48. c
24. b	49. b
25. c	50. e

Chapter 5
Reactions in Aqueous Solution

1. Which one of the following is a nonelectrolyte when dissolved in water.

   a) sodium chloride b) sugar c) copper sulfate d) calcium chloride e) ammonia

2. Which of the following solutions will have the lowest electrical conductivity?

   a) 0.1 M $(NH_4)_3PO_4$ b) 0.1 M $BaCl_2$ c) 0.1 M Na_2SO_4 d) 0.1 M $NaNO_3$

   e) 0.1 M K_3PO_4

3. If solutions of the same concentration are prepared with the following substances, which one will have the highest electrical conductivity?

   a) CH_3CO_2H b) CH_3OH c) NaCl d) $CaCl_2$ e) Cl_2

4. The classification of the following reactions **in order** is

$$HCl(g) + NH_3(g) \rightarrow NH_4Cl(s)$$
$$2HgO(s) \rightarrow O_2(g) + 2Hg(\ell)$$
$$HCl(aq) + AgNO_3(aq) \rightarrow AgCl(s) + HNO_3(aq)$$

   a) acid-base, precipitation, and redox respectively.
   b) precipitation, acid-base, and redox respectively.
   c) redox, precipitation, and acid-base respectively.
   d) acid-base, redox, and precipitation respectively.
   e) redox, acid-base, and precipitation respectively.

5. A precipitate will form when an aqueous solution of lead(II) nitrate is added to an aqueous solution of

   a) NH_4NO_3 b) $Mg(NO_3)_2$ c) $NaNO_3$ d) KNO_3 e) NaCl

6. Which of the following is a weak acid?

   a) NH_3 b) HCl c. $HClO_4$ d) CH_3CO_2H e) HNO_3

7. A solution of nitric acid contains which of the following ions in easily measurable quantities?

   a) H^+, N_2^-, O_2^- b) H^+, NO_2^- c) H_2^+, NO_3^- d) H_2^+, $2NO_2^-$ e) H^+, NO_3^-

8. Identify the spectator ion or ions (if any) in the redox reaction of a solution of lead(II) nitrate with zinc metal.

   a) Pb^{2+} b) Zn^{2+} c) NO_3^- d) H^+ and Pb^{2+} e) no spectator ions are present

9. The driving force for the reaction of zinc metal with a solution of lead(II) nitrate is

a) the formation of a precipitate.

b) the formation of a gas.

c) the evolution of a gas.

d) the dissolving of a solid.

e) the transfer of electrons.

10. Which of the following is predicted to be insoluble in water?

a) $NaBr$ b) NH_4Cl c) FeS d) $(NH_4)_2S$ e) K_2CO_3

11. All of the following salts using the nickel(II) ion would be water insoluble **EXCEPT**

a) S^{2-} b) Cl^- c) PO_4^{3-} d) CO_3^{2-} e) $C_2O_4^{2-}$

12. When a solution of sodium chloride and a solution of ammonium nitrate are mixed

a) $NH_4Cl(s)$ forms. b) $NaNO_3(s)$ forms. c) $NaNH_4(s)$ forms.

d) N_2 and O_2 gases are released. e) neither a precipitate nor a gas is formed.

13. If lead(II) nitrate and sodium chloride solutions are mixed, what is its formula of the precipitate formed (if any)?

a) $PbCl_4$ b) $PbCl_2$ c) $NaNO_3$ d) Na_2Pb e) no precipitate is formed.

14. A white solid is either NaCl or $NaNO_3$. If an aqueous solution is prepared, which reagent will allow you to distinguish between the two compounds?

a) H_2SO_4 b) HCl c) $AgNO_3$ d) $(NH_4)_2SO_4$ e) H_2O

15. The solution which results from the reaction NaOH(aq) and HCl(aq) is the same as the result of the reaction of

a) $Pb(NO_3)_2(aq)$ and NaCl(aq). b) KOH(aq) and HCl(aq). c) NaCl(aq) and $AgNO_3(aq)$.

d) $Na_2CO_3(aq)$ and HCl(aq). e) HCl(aq) and $(NH_4)_2SO_4(aq)$.

16. The compound $Fe(OH)_3$ may be formed in the laboratory by

a) the reaction of $Fe_2O_3(s)$ and NaCl(aq).

b) the reaction of $Fe_2O_3(s)$ and HCl(aq).

c) mixing solutions of $FeSO_4$ and NaOH.

d) mixing solutions of $FeCl_3$ and H_2O.

e) mixing solutions of $FeCl_3$ and NaOH.

17. When an aqueous solution of lead(II) nitrate is treated with an aqueous solution of potassium carbonate, one may observe

a) the formation of a precipitate, $PbCO_3$.

b) the formation of a gas, CO_2.

c) both the formation of $PbCO_3$ precipitate and CO_2 gas.

d) the formation of two precipitates, KNO_3 and $PbCO_3$.

e) no reaction.

18. Which pairs of reagents (if any) could be used in an aqueous solution to prepare pure manganese(II) sulfide by a precipitation reaction?

a) $MnCO_3$ and Ag_2S b) $MnCl_2$ and Na_2S c) $MnCl_2$ and Na_2SO_4 d) $MnSO_4$ and PbS

e) none of these reagents will produce pure manganese(II) sulfide

19. Which of these acids will dissociate 100%?

a) HCO_2H b) CH_3CO_2H c) H_2CO_3 d) H_3BO_3 e) HNO_3

20. Which compound below is a base in aqueous solution?

a) CH_4 b) KCl c) NaOH d) HNO_3 e) HCOOH

21. Write and balance the equation for the reaction of nitric acid with solid sodium carbonate.

a) $H_2NO_3(aq) + NaCO_3(s) \rightarrow H_2O(\ell) + CO_2(g) + NaNO_3(aq)$

b) $2HNO_3(aq) + Na_2CO_3(s) \rightarrow H_2(g) + CO_2(g) + Na_2NO_3(aq)$

c) $2HNO_3(aq) + Na_2CO_3(s) \rightarrow H_2O(\ell) + CO_2(g) + 2NaNO_3(aq)$

d) $2HNO_3(aq) + Na_2CO_3(s) \rightarrow H_2(g) + CO_2(g) + 2NaNO_3(aq)$

e) $2HNO_3(aq) + NaCO_3(s) \rightarrow H_2(g) + CO_2(g) + NaNO_3(aq)$

22. Which equation below best represents the balanced, net ionic equation for the reaction of a dilute solution of calcium hydroxide with a solution of hydrochloric acid?

a) $Ca^{2+}(aq) + 2Cl^-(aq) \rightarrow CaCl_2(s)$

b) $Ca^{2+}(aq) + 2Cl^-(aq) \rightarrow CaCl_2(aq)$

c) $2OH^-(aq) + 2Cl^-(aq) \rightarrow H_2O(\ell) + OCl_2(aq)$

d) $Ca(OH)_2(aq) + 2Cl^-(aq) \rightarrow CaCl_2(s) + 2OH^-(aq)$

e) $OH^-(aq) + H^+(aq) \rightarrow H_2O(\ell)$

23. Which equation below best represents the balanced, net ionic equation for the reaction of a solution of barium nitrate with a solution of potassium carbonate?

a) $Ba^{2+}(aq) + CO_3^{2-}(aq) \rightarrow BaCO_3(s)$

b) $K^+(aq) + NO_3^-(aq) \rightarrow KNO_3(s)$

c) $Ba^{2+}(aq) + NO_3^-(aq) \rightarrow K_2CO_3(aq) + Ba^{2+}(s)$

d) $Ba(NO_3)_2(aq) + K^+(aq) \rightarrow KNO_3(s) + Ba^{2+}(aq)$

e) $Ba(NO_3)_2(aq) + CO_3^{2-}(aq) \rightarrow BaCO_3(s) + 2N_2(g) + 3O_2(g)$

24. Chlorine gas is bubbled into a solution of potassium iodide which is colorless. A red color is observed in the solution. Write the balanced net ionic equation for this reaction.

a) $Cl_2(g) + 2I^-(aq) \rightarrow I_2(aq) + 2Cl^-(aq)$

b) $Cl_2(g) + 2K^+(aq) \rightarrow 2K(s) + 2Cl^-(aq)$

c) $2Cl_2(g) + I_2^-(aq) \rightarrow I_2(aq) + 2Cl_2^-(aq)$

d) $2Cl^-(aq) + 2K^+(aq) \rightarrow 2K(s) + Cl_2(aq)$

e) $Cl_2(g) + 2KI(aq) \rightarrow 2K^+(aq) + I_2(aq) + Cl_2^-(aq)$

25. Solutions of barium nitrate and ammonium sulfate are mixed. A white precipitate is observed. Write the balanced net ionic equation.

a) $Ba^{2+}(aq) + SO_4^{2-}(aq) \rightarrow BaSO_4(s)$

b) $2NH_4^+(aq) + SO_4^{2-}(aq) \rightarrow (NH_4)_2SO_4(s)$

c) $NH_4^+(aq) + NO_3^-(aq) \rightarrow NH_4NO_3(s)$

d) $Ba^{2+}(aq) + 2NH_4^+(aq) + 6H_2O(\ell) \rightarrow Ba(NO_3)_2(s) + 20\ H^+(aq)$

e) $Ba^{2+}(aq) + 2NH_4^+(aq) + 2SO_4^{2-}(aq) \rightarrow (NH_4)_2SO_4(s) + BaSO_4(s)$

26. A solution of sodium carbonate is treated with a solution of nitric acid. Bubbles are observed in the colorless solution. The balanced equation is

a) $Na_2CO_3(aq) + 2HNO_3(aq) \rightarrow H_2O(\ell) + CO_2(g) + 2NaNO_3(aq)$

b) $Na_2CO_3(aq) + 2HNO_3(aq) \rightarrow H_2CO_3(aq) + 2NaNO_3(aq)$

c) $Na_2CO_3(aq) + 2HNO_3(aq) \rightarrow H_2(g) + CO_2(g) + 3\ O_2(g) + N_2(g) + Na_2O(aq)$

d) $2Na_2CO_3(aq) + 2HNO_3(aq) \rightarrow H_2O(\ell) + 2CO(g) + 3\ O_2(g) + NaNO_3(aq)$

e) $2Na_2CO_3(aq) + 2HNO_3(aq) \rightarrow H_2O(\ell) + 2CO_2(g) + N_2(g) + 2NaNO_3(aq)$

27. Consider the equation -- $2NaI(aq) + Cl_2(g) \rightarrow I_2(aq) + 2NaCl(aq)$. The species undergoing reduction is

a) sodium. b) iodide. c) chlorine. d) iodine. e) water.

28. Consider the equation -- $I^-(aq) + ClO^-(aq) \rightarrow IO^-(aq) + Cl^-(aq)$. The oxidizing agent is

a) I^- b) ClO^- c) IO^- d) Cl^- e) H_2O

29. The reducing agent (if any) in the following equation is:

$$2Mg(s) + TiCl_4(\ell) \rightarrow Ti(s) + 2MgCl_2(s)$$

a) $Mg(s)$ b) $TiCl_4(\ell)$ c) $Ti(s)$ d) $MgCl_2(s)$ e) not a redox reaction

30. The oxidation number of chromium in Na_2CrO_4 is

a) -2. b) +2. c) +6. d) -6. e) +8.

31. The oxidation number of chlorine in $KClO_3$ is

a) +6. b) +5. c) -1. d) -2. e) +2.

32. The oxidation number of sulfur in $S_2O_3^{2-}$ is

a) +8. b) +6. c) +4. d) +3. e) +2.

33. The oxidation number of nitrogen in N_2O_5 is

a) -3. b) +2. c) +3. d) +5. e) +10.

34. Which of the following equations represents an oxidation-reduction reaction?

1) $H_2SO_3(aq) \rightarrow H_2O(\ell) + SO_2(g)$

2) $Zn(s) + Cu(NO_3)_2(aq) \rightarrow Cu(s) + Zn(NO_3)_2(aq)$

3) $Zn(s) + S(s) \rightarrow ZnS(s)$

a) 1 b) 1 and 2 c) 2 and 3 d) 1 and 3 e) 1, 2, and 3

35. Which of the following equations represents an oxidation-reduction reaction?

1) $3Mg(s) + N_2(g) \rightarrow Mg_3N_2(s)$

2) $H_2CO_3(aq) \rightarrow H_2O(\ell) + CO_2(g)$

3) $Sr(NO_3)_2(aq) + Na_2SO_4(aq) \rightarrow SrSO_4(s) + 2NaNO_3(aq)$

a) 1 b) 2 and 3 c) 1 and 3 d) 1 and 2 e) 1, 2, and 3

36. Which of the following reactions is **NOT** a redox reaction?

a) $2HgO(s) \rightarrow 2Hg(\ell) + O_2(g)$

b) $H_2(g) + Br_2(g) \rightarrow 2HBr(g)$

c) $H_2CO_3(aq) \rightarrow CO_2(g) + H_2O(\ell)$

d) $2HCl(aq) + Zn(s) \rightarrow H_2(g) + ZnCl_2(aq)$

e) $2NaI(aq) + Cl_2(aq) \rightarrow I_2(aq) + 2NaCl(aq)$

37. How many milliliters of 0.123 M NaOH solution contain 25.0 g of NaOH (molar mass = 40.00 g/mol)?

a) 5.08 mL b) 50.8 mL c) 508 mL d) 625 mL e) 5080 mL

38. If 4.00 mL of water is added to 6.00 mL of 0.0250 M $CuSO_4$, what is the concentration of copper(II) sulfate in the diluted solution?

a) 0.0100 M b) 0.0150 M c) 0.0175 M d) 0.0375 M e) 0.0417 M

39. If you need 1.00 L of 0.125 M H_2SO_4, how would you prepare this solution?

a) Dilute 36.0 mL of 1.25 M H_2SO_4 to a volume of 1.00 L

b) Dilute 20.8 mL of 6.00 M H_2SO_4 to a volume of 1.00 L

c) Add 950. mL of water to 50.0 mL of 3.00 M H_2SO_4

d) Add 500. mL of water to 500. mL of 0.500 M H_2SO_4

e) Add 750. mL of water to 250 mL of 0.375 M H_2SO_4

40. What are the ion concentrations in a 0.12 M solution of $BaCl_2$?

a) $[Ba^{2+}]$ = 0.12 M and $[Cl^-]$ = 0.12 M

b) $[Ba^{2+}]$ = 0.12 M and $[Cl^-]$ = 0.060 M

c) $[Ba^{2+}]$ = 0.12 M and $[Cl^-]$ = 0.24 M

d) $[Ba^{2+}]$ = 0.060 M and $[Cl^-]$ = 0.060 M

e) $[Ba^+]$ = 0.12 M and $[Cl_2^-]$ = 0.12 M

41. What is the ***total*** concentration of ions in a 0.0360 M solution of Na_2CO_3?

a) 0.0120 M b) 0.0720 M c) 0.0900 M d) 0.108 M e) 0.144 M

42. How many grams of Na_2CO_3 (molar mass = 106.0 g/mol) are required for complete reaction with 25.0 mL of 0.155 M HNO_3?

$$Na_2CO_3(aq) + 2HNO_3(aq) \rightarrow 2NaNO_3(aq) + CO_2(g) + H_2O(\ell)$$

a) 0.122 g b) 0.205 g c) 0.410 g d) 20.5 g e) 205 g

43. An aqueous solution of NaCl gives $H_2(g)$, $Cl_2(g)$, and NaOH when an electrical current is passed through the solution.

$$2NaCl(aq) + 2H_2O(\ell) \rightarrow H_2(g) + Cl_2(g) + 2NaOH(aq)$$

If you begin with 10. liters of 0.15 M NaCl, how many grams of NaOH can be formed?

a) 3.0 g b) 6.0 g c) 30. g d) 60. g e) 120 g

44. In the photographic process, silver bromide is dissolved by adding sodium thiosulfate.

$$AgBr(s) + 2Na_2S_2O_3(aq) \rightarrow Na_3Ag(S_2O_3)_2(aq) + NaBr(aq)$$

If you want to dissolve 0.250 g of AgBr (molar mass = 187.8 g/mol), how many milliliters of 0.0138 M $Na_2S_2O_3$ should you add?

a) 96.5 mL b) 193 mL c) 250 mL d) 386 mL e) 425 mL

45. How many milliliters of 0.812 M HCl would be required to titrate 1.33 g of NaOH to the equivalence point?

$$NaOH(aq) + HCl(aq) \rightarrow NaCl(aq) + H_2O(\ell)$$

a) 20.7 mL b) 25.6 mL c) 34.2 mL d) 40.9 mL e) 45.7 mL

46. A soft drink contains an unknown amount of citric acid, $C_6H_8O_7$. If 100. mL of the soft drink require 33.51 mL of 0.0102 M NaOH to neutralize completely the citric acid, how many grams of citric acid (molar mass = 192.13 g/mol) does the soft drink contain per 100 mL? The reaction of citric acid and NaOH is

$$C_6H_8O_7(aq) + 3NaOH(aq) \rightarrow Na_3C_6H_5O_7(aq) + 3H_2O(\ell)$$

a) 0.0219 g b) 0.0660 g c) 0.219 g d) 0.657 g e) 0.782 g

47. What volume of 0.166 M NaOH is needed to neutralize 10.0 mL of 0.261 M HNO_3?

a) 0.0157 mL b) 1.57 mL c) 15.7 mL d) 157 mL e) 1.57 L

48. If 18.5 mL of a NaOH solution is required to neutralize 25.0 mL of 0.457 M HCl, what is the molarity of the NaOH?

a) 8.45×10^{-3} M b) 1.14×10^{-2} M c) 0.618 M d) 1.24 M e) 1.82 M

49. What volume of 0.150 M NaOH is needed to react completely with 3.45 g iodine according to the equation:

$$3\ I_2 + 6NaOH \rightarrow 5NaI + NaIO_3 + 3H_2O\ ?$$

a) 181 mL b) 45.3 mL c) 4.08 mL d) 2.04 mL e) 1.02 mL

50. A 4.00 M solution of H_3PO_4 will contain _____ g of H_3PO_4 in 0.250 L of solution.

a) 196 b) 98.0 c) 49.0 d) 24.0 e) 12.0

Chapter 5: Answers:

1. b	26. a
2. d	27. c
3. d	28. b
4. d	29. a
5. e	30. c
6. d	31. b
7. e	32. e
8. c	33. d
9. e	34. c
10. c	35. a
11. b	36. c
12. e	37. e
13. b	38. b
14. c	39. b
15. d	40. c
16. e	41. d
17. a	42. b
18. b	43. d
19. e	44. b
20. c	45. d
21. c	46. a
22. e	47. c
23. a	48. c
24. a	49. a
25. a	50. b

Chapter 6
Principles of Reactivity: Energy and Chemical Reactions

1. How many calories are equivalent to 364 J?

   a. 1523 cal b. 87.0 cal c. 14.6 cal d. 1.33 cal e. 0.364 cal

2. How many joules are equivalent to 37.7 cal?

   a. 9.01 J b. 9.43 J c. 1.51 J d. 4.184 J e. 158 J

3. The quantity of heat that is needed to raise the temperature of a sample of a substance 1.00 kelvin is called its

   a. heat capacity. b. specific heat capacity. c. enthalpy. d. calorimetry. e. kinetic energy.

4. Specific heat capacity is

   a. amount of heat energy needed to change 1.00 g of substance by 1.00 K.
   b. amount of heat energy needed to change 1.00 mol of substance by 1.00 K.
   c. amount of energy required to melt 1.00 g of substance.
   d. amount of substance that is heated by 1.00 K.
   e. the number of kelvins that 1.00 g of substance is raised by heating it for 1.00 minute.

5. Equal masses of two substances, A and B, each absorb 25 joules of energy. If the temperature of A increases by 4 degrees and the temperature of B increases by 8 degrees, one can say that

   a. the specific heat of A is double that of B.
   b. the specific heat of B is double that of A.
   c. the specific heat of B is negative.
   d. the specific heat of A is negative.
   e. the specific heat of B is triple that of A.

6. Given the following specific heats of metal?

Metal	Specific Heat (J/g°C)
Manganese	0.477
Sodium	1.225
Strontium	0.301
Aluminum	0.899
Beryllium	1.823

If the same amount of heat is added to 100.0 g samples of each of the metals which are all at the same temperature, which metal will have the lowest temperature?

a. Mn b. Na c. Sr d. Al e. Be

7. If 25 J are required to change the temperature of 5.0 g of substance A by 2.0 K, what is the specific heat of substance A?

a. 250 J/g·K b. 63 J/g·K c. 10. J/g·K d. 2.5 J/g·K e. 0.40 J/g·K

8. When 80. J of energy is absorbed by 0.50 mol of water, how much does the temperature rise? The specific heat of water is 4.184 J/g·K.

a. 0.42 K b. 2.1 K c. 3.2 K d. 38 K e. 170 K

9. When 75.4 J of energy is absorbed by 0.250 mole of CCl_4, what is the change in temperature? The specific heat of CCl_4 is 0.861 J/g·K.

a. -17.8 K b. -21.9 K c. 2.28 K d. 9.12 K e. 44.6 K

10. When 0.250 mol of gold is cooled from 50.0°C to 20.0°C, what is the energy transferred? The specific heat of gold is 0.128 J/g·K.

a. +0.96 J b. +189 J c. +756 J d. -189 J e. -756 J

11. The specific heats of three elements are given. Rank the elements in order of their specific heat capacities.

Metal	Specific Heat (J/g·K)
Copper	0.385
Magnesium	1.02
Lead	0.129

a. Mg < Cu < Pb b. Mg < Pb < Cu c. Cu < Pb < Mg d. Cu < Mg < Pb e. Pb < Cu < Mg

12. Calculate the specific heat of a certain unknown metal, if it requires 195 joules of heat to raise the temperature of 12.1 grams of this metal by 34.6°C.

a. 0.335 J/g·K b. 0.466 J/g·K c. 0.714 J/g·K d. 0.816 J/g·K e. 0.928 J/g·K

13. Consider 10.0 g Al, 10.0 g Cu and 10.0 g ethanol (C_2H_5OH). Which requires more energy for a 5.0°C temperature rise than does **1.0** gram of water?

Substance	Specific Heat (J/g·K
Ethanol (C_2H_5OH)	2.46
Water	4.184
Copper	0.385
Aluminum	0.902

a. 10.0 g Al b. 10.0 g Cu c. 10.0 g C_2H_5OH d. both 10.0 g Al and 10.0 g C_2H_5OH

e. 10.0 g Al, 10.0 g Cu and 10.0 g C_2H_5OH

14. How many grams of lead will absorb the same amount of energy as 15.0 g Ag when each metal is heated from 20.0°C to 35.0°C?

Substance	Specific Heat (J/g·K)
Lead	0.129
Silver	0.237

a. 6.50 g b. 27.6 g c. 53.3 g d. 97.9 g e. 225 g

15. How much energy is required to change the temperature of 2.00 g of aluminum from 20.0°C to 25.0°C? The specific heat of aluminum is 0.902 J/g·K.

a. 2.3 J b. 9.0 J c. 0.36 J d. 0.090 J e. 7.6 J

16. When 15.0 grams of an alloy is heated from 20.0°C to 40.0°C it absorbs 727 joules of energy. The specific heat of the alloy is

a. 2.42 J/g·K b. 0.218 J/g·K c. -2.42 J/g·K d. -0.218 J/g·K e. 0.206 J/g·K

17. When 17.0 grams of an alloy is cooled from 35.0°C to 10.0°C it releases 862 joules of energy. The specific heat of the alloy is

a. 0.493 J/g·K b. 0.690 J/g·K c. 2.03 J/g·K d. 3.66 J/g·K e. 50.7 J/g·K

18. If 15.0 g water at 28.0°C is added to 125.0 g water at 20.0°C, what is the final temperature of the resulting mixture?

a. 20.9°C b. 22.6°C c. 23.1°C d. 24.0°C e. 27.3°C

19. If 12.0 g water at 35.0°C is added to 150.0 g water at 22.0°C, what is the final temperature of the resulting mixture?

a. 22.1°C   b. 23.0°C   c. 25.2°C   d. 28.5°C   e. 31.7°C

20. When 115 grams of water at 22.0°C is mixed with an unknown mass of water at a temperature of 58.0°C, the final temperature of the resulting mixture is 45.0°C. What was the mass of the second sample of water?

a. 142 g   b. 187 g   c. 203 g   d. 265 g   e. 289 g

21. When 325 grams of water at 21.0°C is mixed with an unknown mass of water at a temperature of 45.0°C, the final temperature of the resulting mixture is 36.0°C. What was the mass of the second sample of water?

a. 7.24 g   b. 226 g   c. 542 g   d. 874 g   e. 2266 g

22. Calculate the amount of heat needed to change 25.0 g ice at -15.0°C to steam at 100°C. (Some constants for H_2O: Heat of fusion = 333 J/g; heat of vaporization = 2260 J/g; specific heats: Ice = 2.1 J/g·K, water = 4.2 J/g·K, steam 2.0 J/g·K)

a. 76 kJ   b. 65 kJ   c. 48 kJ   d. 26 kJ   e. 11 kJ

23. The standard state of an element or compound is determined at a pressure of ________ and a temperature of ________.

a. 760 atm, 0°C   b. 1 mmHg, 273°C   c. 760 mmHg, 273 K   d. 760 atm, O K   e. 1 atm, 298 K

24. When a piece of aluminum weighing 35.7 grams, and at a temperature of 81.9°C, is placed in a calorimeter containing 75.0 grams of water at 24.9°C, the temperature increases to 28.3°C. If the specific heat of the water is 4.18 J/g·K and the specific heat of the aluminum is 0.902 J/g·K, what is the heat capacity of the calorimeter?

a. 11.6 J/K   b. 194 J/K   c. 496 J/K   d. 660 J/K   e. 2290 J/K

25. For a particular process q = 30 kJ and w = -25 kJ. What conclusions may be drawn for the process?

a. ΔE = 55 kJ

b. ΔE = -55 kJ

c. the transfer of heat energy is from the system to the surroundings.

d. the transfer of work energy is from the surroundings to the system.

e. the system does work on the surroundings.

26. What is ΔE for a system which has the following two steps:

Step 1: The system absorbs 60 J of heat while 40 J of work are performed on it.

Step 2: The system releases 30 J of heat while doing 70 J of work.

a. 110 J b. 100 J c. 90 J d. 30 J e. zero

27. Consider the thermal energy transfer during a chemical process. When heat is transferred to the system, the process is said to be _______ and the sign of q is _______.

a. exothermic, positive b. exothermic, negative c. endothermic, positive

d. endothermic, negative e. enthalpic, negative

28. For the general reaction $2\ A + B_2 \rightarrow 2\ AB$, ΔH is +50.0 kJ. We can conclude that

a. the reaction is endothermic.

b. the surroundings absorb energy.

c. the standard enthalpy of formation of AB is -50.0 kJ.

d. the bond energy of each A-B bond is 50.0 kJ.

e. the molecule AB contains less energy than A or B_2.

29. When two solutions are mixed, the container "feels hot." Thus,

a. the reaction is endothermic.

b. the reaction is exothermic.

c. the energy of the universe is increased.

d. the energy of both the system and the surroundings is decreased.

e. the energy of the system is increased.

30. Which response lists the processes that are endothermic and none that are exothermic?

1. Evaporation of water
2. Sublimation of ice
3. Condensation of steam
4. Freezing of water

a. 1 and 2 only b. 1 and 3 only c. 2 and 3 only d. 3 and 4 only e. 2, 3, and 4 only

31. Which process involves the largest energy change for one mole of 1,4 dichlorobenzene, moth balls?

a. $\Delta H_{vaporization}$ b. $\Delta H_{sublimation}$ c. $\Delta H_{boiling}$ d. ΔH_{fusion} e. $\Delta H_{condensation}$

32. The equation for the standard enthalpy of formation of N_2O_3 is

a. $N_2O(g) + O_2(g) \rightarrow N_2O_3(g)$ b. $N_2O_5(g) \rightarrow N_2O_3(g) + O_2(g)$ c. $NO(g) + NO_2(g) \rightarrow N_2O_3(g)$

d. $N_2(g) + 3/2\ O_2(g) \rightarrow N_2O_3(g)$ e. $2NO(g) + 1/2\ O_2(g) \rightarrow N_2O_3(g)$

33. The equation for the standard enthalpy of formation of hydrazine, N_2H_4, is

a. $2N_2H_4(\ell) \rightarrow 2NH_3(g) + H_2(g)$ b. $2NH_3(g) + H_2(g) \rightarrow N_2H_4(\ell)$

c. $N_2(g) + 2H_2O(\ell) \rightarrow N_2H_4(\ell) + O_2(g)$ d. $N_2(g) + 2H_2(g) \rightarrow N_2H_4(\ell)$

e. $2NO_2(g) + 6H_2(g) \rightarrow N_2H_4(\ell) + 4H_2O(\ell)$

34. The equation for the standard enthalpy of formation for magnesium nitrate $Mg(NO_3)_2(s)$ is

a. $Mg(s) + 2NO_3(g) \rightarrow Mg(NO_3)_2(s)$ b. $Mg(s) + N_2(g) + 3O_2(g) \rightarrow Mg(NO_3)_2(s)$

c. $MgO_2(s) + 2NO_2(g) \rightarrow Mg(NO_3)_2(s)$ d. $2MgO(s) + 2N_2(g) + 5O_2(g) \rightarrow 2Mg(NO_3)_2(s)$

e. $Mg_3N_2(s) + 6NO_2(g) + 3O_2(g) \rightarrow 3Mg(NO_3)_2(s)$

35. Which of the following would have an enthalpy of formation value (ΔH_f) of zero?

a. $H_2O(g)$ b. $O(g)$ c. $H_2O(\ell)$ d. $O_2(g)$ e. $H(g)$

36. Which of the following equations represents an enthalpy change at 25°C and 1 atm that is equal to $\Delta H°_f$?

a. $CO_2(g) + H_2(g) \rightarrow HCOOH(\ell)$ b. $CO(g) + H_2O(\ell) \rightarrow HCOOH(\ell)$

c. $2C(s) + 2H_2(g) + 2O_2(g) \rightarrow 2HCOOH(\ell)$ d. $H_2O(\ell) + C(s) + 1/2\ O_2(g) \rightarrow HCOOH(\ell)$

e. $C(s) + H_2(g) + O_2(g) \rightarrow HCOOH(\ell)$

37. Which equation represents the standard enthalpy of formation for acrylonitrile, C_3H_3N?

a. $3\ C(\text{graphite}) + 3/2\ H_2(g) + 1/2\ N_2(g) \rightarrow C_3H_3N(\ell)$

b. $3\ C(\text{graphite}) + NH_3(g) \rightarrow C_3H_3N(\ell)$

c. $3\ CH_4(g) + NH_3(g) \rightarrow C_3H_3N(\ell) + 6H_2(g)$

d. $3\ CO_2(g) + N_2(g) + 3\ H_2O(g) \rightarrow C_3H_3N(\ell) + NH_3(g) + 9/2\ O_2(g)$

e. $3\ CO_2(g) + N_2(g) + 3\ H_2(g) \rightarrow C_3H_3N(\ell) + NH_3(g) + 3\ O_2(g)$

38. The standard enthalpy of formation of acetylene is represented by the equation

$$H_2(g) + 2C(s) \rightarrow C_2H_2(g) \qquad \Delta H_f° = 227\ \text{kJ/mol}$$

From this information, we can conclude that acetylene

1. will release energy when prepared from its elements.
2. is energetically unstable with respect to its elements.
3. will decompose rapidly into C and H_2.

a. 1 only b. 2 only c. 3 only d. 1 and 2 only e. 2 and 3 only

39. Given the heats of the following reactions:

$2ClF(g) + O_2(g) \rightarrow Cl_2O(g) + F_2O(g)$ $\Delta H = 167.4$ kJ

$2ClF_3(g) + 2O_2(g) \rightarrow Cl_2O(g) + 3F_2O(g)$ $\Delta H = 341.4$ kJ

$2F_2(g) + O_2(g) \rightarrow 2F_2O(g)$ $\Delta H = -43.4$ kJ

Calculate the heat of the reaction of chlorine monofluoride with F_2 according to the equation:

$ClF(g) + F_2(g) \rightarrow ClF_3(g)$

a. -217.5 kJ b. -130.2 kJ c. -108.7 kJ d. 130.2 kJ e. 217.5 kJ

40. Calculate the heat of vaporization for titanium (IV) chloride

$TiCl_4(\ell) \rightarrow TiCl_4(g)$

using the following enthalpies of reaction:

$Ti(s) + 2Cl_2(g) \rightarrow TiCl_4(\ell)$ $\Delta H^\circ = -804.2$ kJ

$TiCl_4(g) \rightarrow 2Cl_2(g) + Ti(s)$ $\Delta H^\circ = 763.2$ kJ

a. -1567 kJ b. -783.7 kJ c. -41 kJ d. 41 kJ e. 1165 kJ

41. Calculate the enthalpy of reaction for the process

$D + F \rightarrow G + M$

using the following equations and data:

$G + C \rightarrow A + B$ $\Delta H^\circ = +277$ kJ

$C + F \rightarrow A$ $\Delta H^\circ = +303$ kJ

$D \rightarrow B + M$ $\Delta H^\circ = -158$ kJ

a. -132 kJ b. +422 kJ c. +132 kJ d. -184 kJ e. -422 kJ

42. Calculate the enthalpy of combustion of C_3H_6 based on the equation

$C_3H_6(g) + 9/2\ O_2(g) \rightarrow 3CO_2(g) + 3H_2O$

using the following data:

$3C(s) + 3H_2(g) \rightarrow C_3H_6(g)$ $\Delta H^\circ = 53.3$ kJ

$C(s) + O_2(g) \rightarrow CO_2(g)$ $\Delta H^\circ = -394$ kJ

$H_2(g) + 1/2\ O_2(g) \rightarrow H_2O(\ell)$ $\Delta H^\circ = -286$ kJ

a. -1517 kJ b. -733 kJ c. -626 kJ d. 1304 kJ e. 2093 kJ

43. Calculate the enthalpy for the decomposition of calcium carbonate according to the equation

$$CaCO_3(s) \rightarrow CaO(s) + CO_2(g)$$

using the following:

$Ca(s) + 1/2\ O_2(g) \rightarrow CaO(s)$ $\Delta H° = -635$ kJ

$C(s) + O_2(g) \rightarrow CO_2(g)$ $\Delta H° = -394$ kJ

$Ca(s) + C(s) + 3/2\ O_2(g) \rightarrow CaCO_3(s)$ $\Delta H° = -1207$ kJ

a. 1447 kJ b. 966 kJ c. 178 kJ d. -178 kJ e. -1447 kJ

44. Calculate the standard enthalpy change for the reaction

$$C_2H_2(g) + H_2(g) \rightarrow C_2H_4(g)$$

based on the following standard enthalpies of formation:

$\Delta H°_f[C_2H_2(g)] = +226.7$ kJ/mol

$\Delta H°_f[C_2H_4(g)] = +52.3$ kJ/mol

a. 174.4 kJ b. -56.4 kJ c. -174.4 kJ d. -279.0 kJ e. -321.1 kJ

45. Calculate the enthalpy of reaction for the process

$$NO_2(g) + CO(g) \rightarrow CO_2(g) + NO(g)$$

using the standard enthalpies of formation:

NO_2 = 34 kJ/mol; CO = -111 kJ/mol; CO_2 = -394 kJ/mol; NO = 90 kJ/mol

a. 381 kJ b. 339 kJ c. 227 kJ d. -227 kJ e. -339 kJ

46. Calculate the standard enthalpy of reaction for the process

$$3NO(g) \rightarrow N_2O(g) + NO_2(g)$$

using the standard enthalpies of formation:

NO = 90 kJ/mol; N_2O = 82.1 kJ/mol; NO_2 = 34.0 kJ/mol

a. -153.9 kJ b. -26.1 kJ c. 26.1 kJ d. 206 kJ e. 386 kJ

47. Using the following information

$C(s) + 2Cl_2(g) \rightarrow CCl_4(\ell)$ $\Delta H° = -135.4$ kJ

$H_2(g) + Cl_2(g) \rightarrow 2HCl(g)$ $\Delta H° = -184.6$ kJ

$2H_2(g) + C(s) \rightarrow CH_4(g)$ $\Delta H° = -74.8$ kJ

calculate the standard enthalpy of reaction for the process:

$CH_4(g) + 4Cl_2(g) \rightarrow CCl_4(\ell) + 4HCl(g)$ $\Delta H°_{rxn} = ?$

a. -429.8 kJ b. 152.9 kJ c. 302.1 kJ d. 394.4 kJ e. 579.4 kJ

48. The standard molar enthalpy change is -802.3 kJ for the combustion of methane gas.

$$CH_4(g) + 2\ O_2(g) \rightarrow CO_2(g) + 2H_2O(g)$$

Calculate the standard molar enthalpy of formation for methane based on the following standard enthalpies of formation:

$\Delta H^\circ_f[CO_2(g)] = -393.5$ kJ/mol

$\Delta H^\circ_f[H_2O(g)] = -241.8$ kJ/mol

a. -1679 kJ/mol b. -125.4 kJ/mol c. -74.8 kJ/mol d. 74.8 kJ/mol e. 892.4 kJ/mol

49. The standard molar enthalpy change is -905.2 kJ for the oxidation of ammonia.

$$4NH_3(g) + 5\ O_2(g) \rightarrow 4\ NO(g) + 6H_2O(g)$$

Calculate the standard molar enthalpy of formation for ammonia based on the following standard enthalpies of formation:

$\Delta H^\circ_f[NO(g)] = +90.3$ kJ/mol

$\Delta H^\circ_f[H_2O(g)] = -241.8$ kJ/mol

a. -46.1 kJ/mol b. -92.2 kJ/mol c. -184.4 kJ/mol d. -226.7 kJ/mol e. -498.8 kJ/mol

50. The standard molar enthalpy change is -1277.3 kJ for the combustion of ethanol

$$C_2H_5OH(g) + 3\ O_2(g) \rightarrow 2CO_2(g) + 3H_2O(g)$$

Calculate the standard molar enthalpy of formation for ethanol based on the following standard enthalpies of formation:

$\Delta H^\circ_f[CO_2(g)] = -393.5$ kJ/mol

$\Delta H^\circ_f[H_2O(g)] = -241.8$ kJ/mol

a. -642.7 kJ/mol b. -235.1 kJ/mol c. -122.9 kJ/mol d. 235.1 kJ/mol e. 642.7 kJ/mol

Chapter 6: Answers:

1. b	26. e
2. e	27. c
3. a	28. a
4. a	29. b
5. a	30. a
6. e	31. b
7. d	32. d
8. b	33. d
9. c	34. b
10. d	35. d
11. d	36. e
12. b	37. a
13. d	38. b
14. b	39. c
15. b	40. d
16. a	41. a
17. c	42. e
18. a	43. c
19. b	44. c
20. c	45. d
21. c	46. a
22. a	47. a
23. e	48. c
24. b	49. a
25. e	50. b

Chapter 7
Atomic Structure

1. Assume that the nucleus of an atom has a radius of 1.00 cm. What is the most reasonable value for the radius of the entire atom?

   a. 5.00 cm b. 5.00 μm c. 500 nm d. 500 pm e. 1.00×10^5 cm

2. Which of the following produces radiation of the highest frequency?

   a. x-rays b. AM radio c. FM radio d. microwave oven e. radar

3. Which of the following types of radiation has the lowest energy?

   a. gamma rays b. visible c. ultraviolet d. infrared e. radio

4. Which of the following types of radiation has the longest wavelength?

   a. gamma rays b. visible c. ultraviolet d. radar e. x-ray

5. Which of the following has the longest wavelength?

   a. blue light b. red light c. yellow light d. green light e. orange light

6. What is the frequency of yellow light having a wavelength of 562 nanometers?

   a. $5.34 \times 10^{14} s^{-1}$ b. $5.34 \times 10^{5} s^{-1}$ c. $1.87 \times 10^{-6} s^{-1}$ d. $1.87 \times 10^{-15} s^{-1}$ e. $1.18 \times 10^{-27} s^{-1}$

7. What is the frequency of ultraviolet radiation having a wavelength of 46.3 nanometers?

   a. $1.54 \times 10^{-16} s^{-1}$ b. $1.54 \times 10^{-9} s^{-1}$ c. $1.18 \times 10^{-7} s^{-1}$ d. $6.48 \times 10^{6} s^{-1}$ e. $6.48 \times 10^{15} s^{-1}$

8. If a radio station has a frequency of 90.3 megahertz (MHz), what is the wavelength of the station in cm? (1 MHz = 1.00×10^6 cycles/second)

   a. 0.254 cm b. 0.369 cm c. 110 cm d. 271 cm e. 332 cm

9. If the frequency is observed to be 1.00×10^{12} Hz for a microwave signal, what is the wavelength of this radiation in centimeters? (1 Hz = 1 cycle per second)

   a. 3.00×10^{-4} cm b. 3.00×10^{-2} cm c. 3.34 cm d. 3000 cm e. 3340 cm

10. If red light has a frequency of 4.28×10^{14} Hz, what is the wavelength of this light?

    a. 0.650 nm b. 2.10 nm c. 65.0 nm d. 650. nm e. 700. nm

11. The green light associated with the aurora borealis is emitted by excited oxygen atoms. The frequency of this light is 5.38×10^{14} s^{-1}. What is the wavelength of this light in nm? ($c = 3.00 \times 10^8$ m/s)

a. 557 nm b. 610 nm c. 649 nm d. 6100 nm e. 6490 nm

12. If wavelength of ultraviolet light is 105 nanometers, what is the energy of one quantum of this radiation?

a. 3.21×10^{-41} J b. 8.76×10^{-29} J c. 2.15×10^{-27} J d. 1.15×10^{-23} J e. 1.89×10^{-18} J

13. If the wavelength of blue light is 412 nanometers, what is the energy of one quantum of this radiation?

a. 8.36×10^{-29} J b. 5.62×10^{-23} J c. 4.82×10^{-19} J d. 2.91×10^{-15} J e. 7.31×10^{-14} J

14. What is the energy of a mole of photons of orange light with a wavelength of 585 nanometers?

a. 1.20×10^{-52} J/mol b. 7.41×10^{-29} J/mol c. 1.61×10^{-27} J/mol d. 2.78×10^{-18} J/mol e. 2.05×10^{5} J/mol

15. What is the energy of a mole of photons of infrared radiation of wavelength 1.72×10^{-3} cm?

a. 6.96×10^{3} J/mol b. 1.90×10^{5} J/mol c. 2.83×10^{11} J/mol d. 1.15×10^{20} J/mol e. 1.04×10^{37} J/mol

16. Planck suggested that all energy gained or lost by an atom must be some integral multiple of a minimum amount of energy called a(n)

a. electron. b. spectrum. c. magnetic moment. d. quantum. e. orbital.

17. In the photoelectric effect, no electrons are emitted from the surface of a silver foil when the frequency of the incident light is less than 1.15×10^{15} Hz. At frequencies > 1.15×10^{15} Hz electrons were emitted. What is the minimum energy necessary to eject an electron from the silver? (1 Hz = 1 cycle/second = 1 s^{-1})

a. 6.63×10^{-34} J b. 1.26×10^{-22} J c. 7.62×10^{-19} J d. 3.44×10^{23} J e. 1.74×10^{48} J

18. According to the experiments concerned with the photoelectric effect, what was the result of increasing the intensity of the light striking the metal surface?

a. The number of electrons emitted was increased.
b. The energy of the electrons emitted was increased.
c. Both the number and energy of the electrons was increased.
d. Both the number and energy of the electrons was decreased.
e. There was no change in the number or energy of the electrons.

19. What is the binding energy of an electron in a photosensitive metal in J/mol if light of the frequency of 6.0×10^{14} s^{-1} impinging on the metal surface is just able to eject electrons?

a. 6.6×10^{-43} J/mol b. 4.0×10^{-19} J/mol c. 2.4×10^{5} J/mol d. 4.3×10^{8} J/mol

e. 7.2×10^{9} J/mol

20. The mathematic expression(s) which correctly give(s) the relationship(s) between the speed, wavelength, and frequency of electromagnetic radiation is (are)

1. $\nu = \frac{c}{\lambda}$

2. $c = \nu\lambda$

3. $\lambda = \frac{c}{\nu}$

a. 1 only b. 2 only c. 3 only d. 1 and 2 only e. 1, 2, and 3

21. When ignited, a barium compound burns with a green flame. The wavelength of this light is less than that of

a. X-rays. b. gamma rays. c. ultraviolet light. d. violet light. e. orange light.

22. Which of the following best supports the concept that electrons in atoms have quantized energies?

a. Bohr theory

b. Ionization energy of hydrogen

c. Emission spectrum of mercury

d. The alpha particle scattering experiment

e. The photoelectric experiment

23. For a particular element, a photon of yellow light of wavelength of 585 nm resulted when an electron fell from the third energy level to the second energy level. From this information we can determine

a. the energy of the n = 2 level.

b. the energy of the n = 3 level.

c. the sum of the energies of n = 2 and n = 3.

d. the sum of the energies of n = 1, n = 2, and n = 3.

e. the difference in energies between n = 2 and n = 3.

24. For an electron (mass = 9.109×10^{-31} kg) moving with a velocity of 3.00×10^{7} m/s, what is the de Broglie wavelength in meters?

a. 1.37×10^{9} m b. 6.28×10^{-8} m c. 7.27×10^{-10} m d. 1.37×10^{-10} m e. 2.42×10^{-11} m

25. For a neutron (mass = 1.675×10^{-27} kg) moving with a velocity of 4.20 m/s, what is the de Broglie wavelength in nanometers?

a. 0.245 nm   b. 0.824 nm   c. 1.34 nm   d. 94.2 nm   e. 115 nm

26. Which of the following electronic transitions in a hydrogen atom would have the longest wavelength?

a. n = 4 to n = 1   b. n = 4 to n = 2   c. n = 2 to n = 1   d. n = 4 to n = 3   e. n = 1 to n = 0

27. Which of the following electronic transitions in a hydrogen atom would have the highest energy?

a. n = 4 to n = 1   b. n = 4 to n = 2   c. n = 2 to n = 1   d. n = 4 to n = 3   e. n = 1 to n = 0

28. Which of the following transitions in the hydrogen atom results in the emission of light of the longest wavelength?

a. n = 1 to n = 2   b. n = 3 to n = 1   c. n = 2 to n = 1   d. n = 4 to n = 3   e. n = 1 to n = 4

29. Which of the following transitions in the hydrogen atom results in the emission of light of the shortest wavelength?

a. n = 4 to n = 3   b. n = 1 to n = 2   c. n = 3 to n = 1   d. n = 2 to n = 1   e. n = 4 to n = 1

30. What type of orbital is designated n = 3, ℓ = 2, m_ℓ = 0?

a. 2s   b. 3s   c. 3p   d. 3d   e. 4d

31. What type of orbital is designated n = 4, ℓ = 3, m_ℓ = -1?

a. 3p   b. 3d   c. 4p   d. 4d   e. 4f

32. What type of orbital is designated n = 2, ℓ = 1, m_ℓ = -1?

a. 2p   b. 3p   c. 2s   d. 3s   e. 4p

33. What type of orbital (if any) is designated n = 5, ℓ = 2, m_ℓ = -2?

a. 5p   b. 3p   c. 4d   d. 5d   e. no orbital is identified

34. What type of orbital (if any) is designated n = 3, ℓ = 1, m_ℓ = -2?

a. 3s   b. 3p   c. 2p   d. 3d   e. no orbital is identified

35. What is the maximum number of orbitals possible when ℓ = 1?

a. zero   b. one   c. three   d. five   e. nine

36. What is the maximum number of orbitals possible when $\ell = 3$?

a. zero b. three c. five d. seven e. nine

37. When $\ell = 4$, what set of orbitals is designated?

a. f b. p c. s d. d e. g

38. When $\ell = 3$, what set of orbitals is designated?

a. g b. p c. f d. d e. s

39. What is the maximum number of orbitals that can be identified by the quantum numbers $n = 4, \ell = 3, m_\ell = -2$?

a. 0 b. 1 c. 3 d. 5 e. 7

40. What is the maximum number of orbitals that can be identified by the quantum numbers $n = 3, \ell = 3, m_\ell = -2$?

a. 0 b. 1 c. 3 d. 5 e. 7

41. What name is given to a region of an electron probability density graph where the probability of finding the electron is zero?

a. node b. wave function c. orbital d. lobe e. excited state

42. The lowest-energy stationary state of an atom is called its ___________.

a. wave function b. node c. ground state d. orbital e. permanent state

43. Which of the following sets of quantum numbers is not allowed?

a. $n = 3, \ell = 3, m_\ell = +1$ b. $n = 3, \ell = 1, m_\ell = 0$ c. $n = 3, \ell = 0, m_\ell = 0$

d. $n = 4, \ell = 3, m_\ell = -2$ e. $n = 4, \ell = 2, m_\ell = +2$

44. Which of the following sets of quantum numbers is not allowed?

a. $n = 1, \ell = 0, m_\ell = 0$ b. $n = 2, \ell = 0, m_\ell = 0$ c. $n = 2, \ell = 2, m_\ell = +1$

d. $n = 3, \ell = 1, m_\ell = 0$ e. $n = 3, \ell = 1, m_\ell = +1$

45. When $n = 2$, which of the following is a possible value for ℓ?

a. -2 b. 0 c. + 2 d. 4 e. 8

46. When $\ell = 2$, the possible values of m_ℓ are

a. 0 b. 0, 1, 2 c. +1, 0, -1 d. +2, +1, 0, -1, -2 e. +3, +2, +1, 0, -1, -2, -3

47. According to the Heisenberg's uncertainty principle, if one attempts simultaneously to measure the position and momentum of an electron, the more exactly the position is measured, the greater will be the __________ in the momentum measurement.

a. probability   b. uncertainty   c. certainty   d. polarity   e. energy

48. The quantum number ℓ represents the

a. number of valence electrons.
b. number of orbitals.
c. shape of the orbital.
d. orientation of the orbital.
e. momentum of the electron.

49. The quantum number m_ℓ represents the

a. number of valence electrons.
b. number of orbitals.
c. shape of the orbital.
d. orientation of the orbital.
e. momentum of the electron.

50. According to the Bohr atomic theory, when an electron moves from one energy level to another further from the nucleus

a. energy is absorbed.
b. energy is emitted.
c. light is emitted.
d. photons are discharged.
e. no change in energy is observed.

Chapter 7: Answers:

1. e	26. d
2. a	27. a
3. e	28. d
4. d	29. e
5. b	30. d
6. a	31. e
7. e	32. a
8. e	33. d
9. b	34. e
10. e	35. c
11. a	36. d
12. e	37. e
13. c	38. c
14. e	39. b
15. a	40. a
16. d	41. a
17. c	42. c
18. a	43. a
19. c	44. c
20. e	45. b
21. e	46. d
22. c	47. b
23. e	48. c
24. e	49. d
25. d	50. a

Chapter 8
Atomic Electron Configurations and Chemical Periodicity

1. How many electrons can be described by the set of quantum numbers $n = 3$, $\ell = 1$, $m_\ell = -1$, $m_s = -1/2$?

   a. 18 b. 12 c. 1 d. 0 e. 6

2. How many electrons can be described by the set of quantum numbers $n = 3$, $\ell = 3$, $m_\ell = -1$, $m_s = -1/2$?

   a. 18 b. 6 c. 2 d. 1 e. 0

3. What is the maximum number of orbitals that can be identified by the quantum numbers $n = 4$, $\ell = 3$, $m_\ell = -2$?

   a. 0 b. 1 c. 3 d. 5 e. 7

4. What is the maximum number of electrons that can be placed in a single 5f orbital?

   a. two b. seven c. eight d. fourteen e. eighteen

5. Which of the following elements is paramagnetic?

   a. P b. Mg c. Zn d. Ar e. Ba

6. Which one of the following ions is paramagnetic?

   a. Zn^{2+} b. Ca^{2+} c. Ga^{3+} d. Ga^{+} e. Fe^{3+}

7. Which one of the following ions is paramagnetic?

   a. F^{-} b. O^{2-} c. V^{2+} d. Sn^{2+} e. Ba^{2+}

8. Which of the following elements has three unpaired electrons in a 3+ ion?

   a. aluminum b. iron c. chromium d. scandium e. arsenic

9. An element has the electron configuration $1s^22s^22p^63s^23p^64s^23d^{10}$. If the element forms an ion, what is its charge?

   a. +1 b. +2 c. -2 d. -6 e. 0, no ion is possible.

10. Which of the following is the correct electron configuration for the chromium(III) ion?

    a. [Ar] $4s^23d^4$ b. [Ar] $4s^03d^3$ c. [Ar] $4s^23d^2$ d. [Ar] $4s^23d^6$ e. [Ar] $4s^03d^1$

11. Which of the following is the correct electron configuration for the nitride ion?

a. $1s^22s^22p^3$ b. $1s^22s^22p^4$ c. [He] d. $1s^22s^22p^6$ e. $1s^22s^2$

12. Which of the following particles would be most paramagnetic?

a. P b. Ga c. Br d. Cl^- e. Na^+

13. Which of the following sets of quantum numbers is not allowed?

a. $n = 1, \ell = 0, m_\ell = 0, m_s = -1/2$
b. $n = 2, \ell = 1, m_\ell = +1, m_s = +1/2$
c. $n = 3, \ell = 0, m_\ell = -1, m_s = +1/2$
d. $n = 4, \ell = 2, m_\ell = +1, m_s = -1/2$
e. $n = 5, \ell = 2, m_\ell = 0, m_s = -1/2$

14. Which of the following sets of quantum numbers is not allowed?

a. $n = 1, \ell = 0, m_\ell = 0, m_s = 0$
b. $n = 2, \ell = 1, m_\ell = +1, m_s = +1/2$
c. $n = 2, \ell = 0, m_\ell = 0, m_s = +1/2$
d. $n = 3, \ell = 2, m_\ell = +1, m_s = -1/2$
e. $n = 3, \ell = 2, m_\ell = 0, m_s = -1/2$

15. Which of the following sets of quantum numbers is not allowed?

a. $n = 3, \ell = 2, m_\ell = 0, m_s = -1/2$
b. $n = 3, \ell = 2, m_\ell = +2, m_s = +1/2$
c. $n = 2, \ell = 2, m_\ell = -1, m_s = +1/2$
d. $n = 4, \ell = 2, m_\ell = +1, m_s = -1/2$
e. $n = 4, \ell = 3, m_\ell = -3, m_s = -1/2$

16. Which of the following elements is a p-block element?

a. copper b. chlorine c. chromium d. sodium e. silver

17. Which of the following elements is a d-block element?

a. copper b. chlorine c. aluminum d. sodium e. lead

18. What element has the electron configuration $1s^22s^22p^63s^23p^4$?

a. O b. S c. Se d. Si e. Ge

19. Which rare gas symbol would be used for the rare gas notation for the electronic configuration of the element silver?

a. argon b. neon c. krypton d. xenon e. radon

20. What element has the electron configuration [Kr] $4d^5 5s^1$?

a. W b. Ru c. Mo d. Pm e. Cr

21. What element has the electron configuration [Rn] $5f^3 6d^1 7s^2$?

a. U b. Nd c. Br d. Ta e. Bi

22. What element has the following electron configuration?

[Ar] $\frac{\uparrow}{}$ _ _ _ _ (3 d) $\frac{\uparrow\downarrow}{}$ (4s) _ _ _ (4p)

a. Sc b. Cr c. Ca d. Al e. Mo

23. What is the electronic configuration of Se^{2-}?

a. [Ar] $4s^2 4p^6$ b. [Ar] $4s^2 4p^8$ c. [Ar] $4s^2 4p^6 3d^{10}$ d. [Ar] e. [Kr] $3d^{10}$

24. Which of the following ions is diamagnetic?

a. Ti^{2+} b. Mg^{2+} c. V^{2+} d. Cr^{2+} e. Co^{2+}

25. The number of unpaired electrons in the selenium atom is

a. 0 b. 2 c. 4 d. 6 e. 16

26. How many unpaired electrons are present in Fe^{+2}?

a. 0 b. 2 c. 4 d. 5 e. 6

27. Which of the following atoms has the largest number of valence electrons?

a. Al b. P c. Sr d. Ga e. Ca

28. Which of the following atoms has the smallest first ionization energy?

a. Al b. P c. Sr d. Ga e. Rb

29. Which of the following particles has the lowest 2nd ionization energy?

a. F b. O c. Na d. Mg e. Li

30. Which of the following has the lowest 1st ionization energy?

a. F b. O c. Na d. Mg e. Ne

31. Which of the following particles would be predicted to be paramagnetic?

a. Na b. Ne c. Mg d. O^{2-} e. F^-

32. If metallic character is characterized as the ability to lose electrons easily, the most metallic of the atoms Ba, Mg, Pb, Sn, and Zn is

a. Ba b. Mg c. Pb d. Sn e. Zn

33. Rank Na, Mg, and K in order of increasing 2nd ionization energy.

a. Mg < Na < K b. K < Mg < Na c. Na < K < Mg d. K < Na < Mg e. Mg < K < Na

34. Rank Ba, Ca, Na in order of increasing 2nd ionization energy.

a. Ba < Ca < Na b. Ba < Na < Ca c. Ca < Ba < Na d. Na < Ca < Ba e. Na < Ba < Ca

35. Which of the following elements would have the greatest difference between the first and the second ionization energy?

a. lithium b. carbon c. magnesium d. fluorine e. nitrogen

36. Which of the following elements has the most metallic character?

a. calcium b. lithium c. gallium d. cesium e. potassium

37. Which of the following best describes the variation of ionization energy of the elements with respect to their position on the periodic table?

a. Increases across a period, increases slowly down a group.

b. Increases across a period, increases quickly down a group.

c. Increases across a period, decreases down a group.

d. Decreases across a period, decreases down a group.

e. Decreases across a period, increases down a group.

38. A measure of the ability of a gaseous atom to acquire an electron to become negatively charged is called its

a. ionization energy. b. polarizability. c. electron affinity. d. electronegativity.

e. electron density.

39. Which of the following elements has the highest electron affinity?

a. Cl b. N c. C d. P e. Na

40. What ion has the following electron configuration?

_ (4s) ↑ ↑ ↑ _ _ (3d)

a. Sc^{3+} b. V^{2+} c. Ca^{2+} d. Mn^{5+} e. Fe^{3+}

41. Which of the following particles has the largest radius?

a. He b. F^- c. O^{2-} d. Mg^{2+} e. N^{3-}

42. Which one of the following atoms has the largest atomic radius?

a. Al b. Ge c. Ga d. Si e. P

43. Of the particles K^+, Ca^{2+}, S^{2-}, and Cl^-, which one, if any, has the largest ionic radius?

a. K^+ b. Ca^{2+} c. S^{2-} d. Cl^- e. They are all the same.

44. Which of the following groups of elements is arranged correctly in order of increasing affinity for electrons?

a. Mg < S < Al < Cl b. Al < Mg < S < Cl c. Mg < Al < S < Cl d. Cl < S < Mg < Al
e. S < Cl < Mg < Al

45. Which of the following is expected to have the smallest radius?

a. P^{3-} b. S^{2-} c. Cl^- d. K^+ e. Ca^{2+}

46. What is the group number of the main group elements of the periodic table that are expected to form +2 ions?

a. group 1A b. group 2A c. group 3A d. group 4A e. group 5A

47. Which of the following ions is **least** likely to be formed?

a. Be^{2+} b. S^{2-} c. Al^{3+} d. Na^{2+} e. H^+

48. Which of the following ions is **most** likely to be formed?

a. Mg^{3+} b. Na^+ c. Al^{4+} d. N^{3-} e. B^+

49. When arranged in order of increasing atomic number, the elements exhibit periodicity for all of the following properties **except**

a. 1st ionization energy. b. atomic radii. c. atomic masses. d. electron affinity.

e. 2nd ionization energy.

50. Which of the following most probably cannot exist?

a. $BaBr_2$ b. KBr_2 c. Br_2 d. OBr_2 e. $FeBr_2$

Chapter 8: Answers:

1. c	26. c
2. e	27. b
3. b	28. e
4. a	29. d
5. a	30. c
6. e	31. a
7. c	32. a
8. c	33. e
9. b	34. a
10. b	35. a
11. d	36. d
12. a	37. c
13. c	38. c
14. a	39. a
15. c	40. b
16. b	41. e
17. a	42. c
18. b	43. c
19. c	44. c
20. c	45. e
21. a	46. b
22. a	47. d
23. c	48. b
24. b	49. c
25. b	50. b

Chapter 9
Bonding and Molecular Structure; Fundamental Concepts

1. Which species has more than eight electrons around the central atom?

a. BF_3 b. BF_4^- c. BrF_3 d. PF_3 e. OF_2

2. Which of the following compounds exhibits ionic bonding?

a. CCl_4 b. $MgCl_2$ c. Cl_2 d. PCl_3 e. OF_2

3. Which of the following elements is most likely to form compounds involving an expanded valence shell of electrons?

a. Li b. N c. F d. Ne e. S

4. Which of the following element combinations is likely to produce ionic bonds in a compound?

a. lithium and fluorine b. boron and oxygen c. nitrogen and oxygen
d. phosphorus and sulfur e. chlorine and bromine

5. Which of the following element combinations is likely to produce covalent bonds in a compound?

a. potassium and fluorine b. magnesium and oxygen c. nitrogen and oxygen
d. sodium and chlorine e. sodium and fluorine

6. Which of the following elements is **most** likely to participate in the formation of multiple bonds?

a. H b. Na c. Cl d. S e. F

7. Which compound has the most ionic bond?

a. LiCl b. LiF c. KF d. KCl e. NaCl

8. Which of the following groups contains **no** ionic compounds?

a. H_2O, MgO, NO_2 b. CO_2, SO_2, H_2S c. CCl_4, $CaCl_2$, HCl d. Na_2S, SO_2, CS_2
e. Mg_3N_2, NCl_3, HOCl

9. Which of the following compounds would be expected to have the highest melting point?

a. LiF b. LiCl c. NaBr d. CsF e. NaCl

10. Which of the following compounds would be expected to have the lowest melting point?

a. NaCl b. LiF c. NaBr d. CsF e. CsI

11. Which of the following salts is expected to have the highest melting point?

a. NaF b. NaCl c. NaI d. NaBr e. KI

12. Which of the following elements is most likely to form compounds involving an expanded valence shell of electrons?

a. O b. Na c. P d. N e. C

13. Which of the following is not a correct Lewis dot structure?

a. 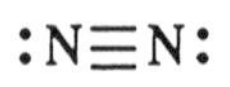 :N≡N:

b. 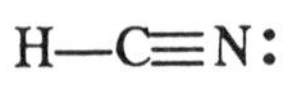 H—C≡N:

c. [:N≡O:]$^-$

d. 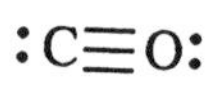 :C≡O:

e. 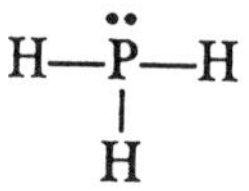

14. Which of the following is not a correct Lewis dot structure?

a. 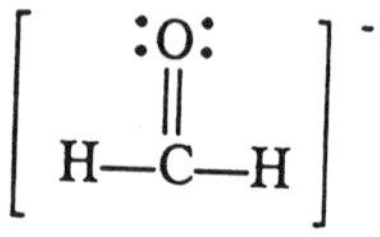

b. 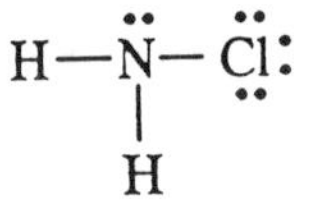

c. 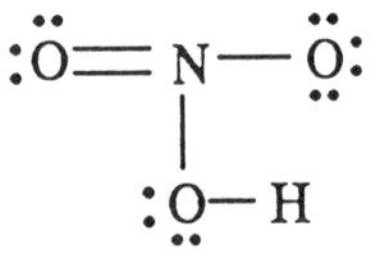

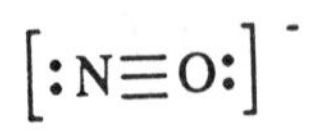

d. 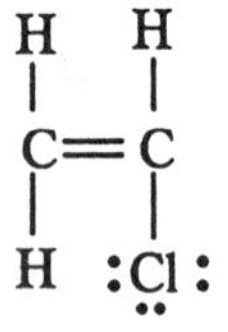

e. 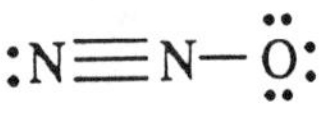

15. Which of the following is (are) CORRECT resonance structure(s) for the formate ion?

1. 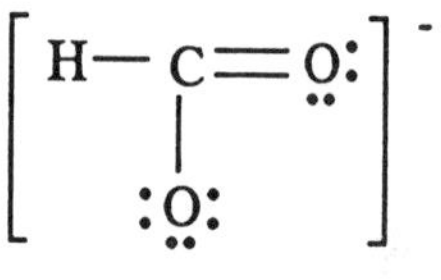

2. 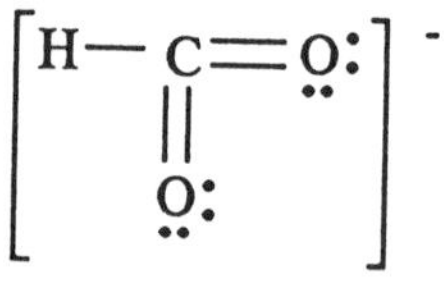

3. 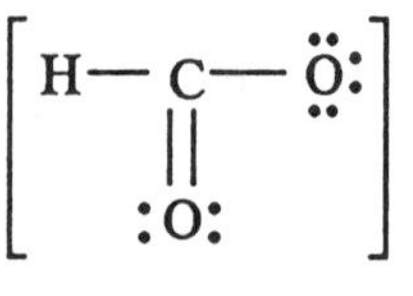

a. 1 only b. 2 only c. 3 only d. 1 and 3 only e. 1, 2, and 3

16. Which of the following is (are) **CORRECT** resonance structure(s) for the N_2O molecule?

1. 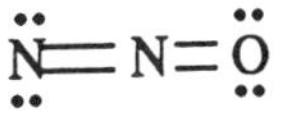

2. 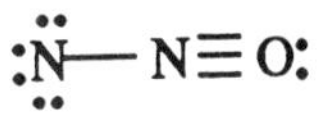

3. 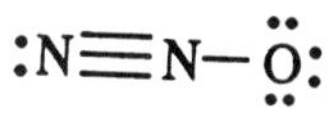

a. 1 only b. 2 only c. 3 only d. 2 and 3 only e. 1, 2, and 3

17. Which of the following is (are) **CORRECT** Lewis dot structures?

1. 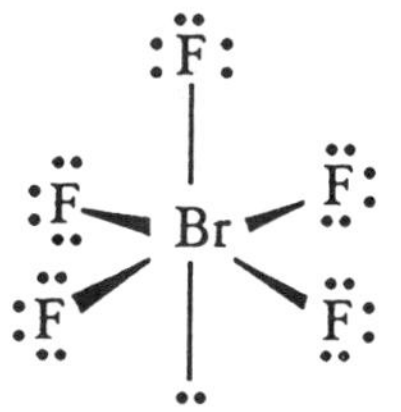

2. 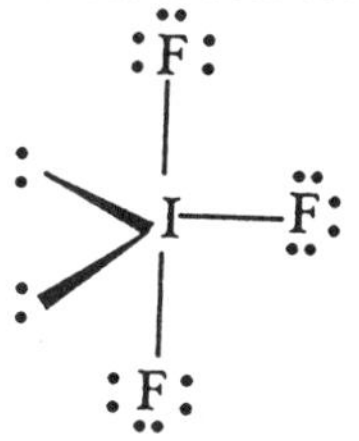

3. 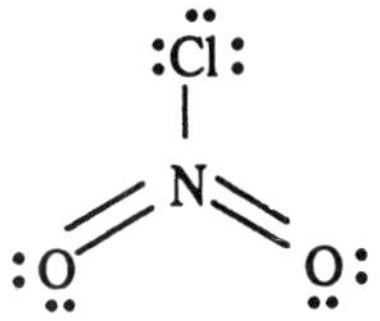

a. 1 only b. 2 only c. 3 only d. 1 and 2 only e. 2 and 3 only

18. The Lewis structure

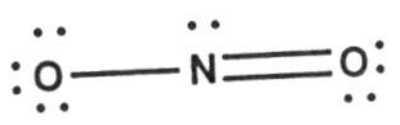

represents

a. NO_2 b. NO_2^- c. NO_2^+ d. both NO_2^- and NO_2^+ e. NO_2, NO_2^-, and NO_2^+

19. The Lewis structure

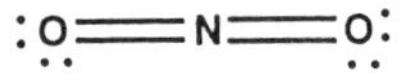

represents

a. NO_2 b. NO_2^- c. NO_2^+ d. both NO_2^- and NO_2^+ e. NO_2, NO_2^-, and NO_2^+

20. How many unshared election pairs (lone pairs) are in a molecule of SO_2?

a. 2 b. 6 c. 7 d. 9 e. 12

21. Which one of the following would have a Lewis structure(s) most like that of CO_3^{2-}?

a. NO_3^- b. NH_3 c. CH_3^+ d. SO_4^{2-} e. SO_3^{2-}

22. Which one of the following species would have a Lewis structure(s) most like that of carbon disulfide, CS_2?

a. NO_2 b. NO_2^- c. NO_2^+ d. HCN e. ICl_2^-

23. Which one of the following has a Lewis structure most like that of SO_3?

a. PO_4^{3-} b. SO_4^{2-} c. SO_3^{2-} d. ClO_3^- e. CO_3^{2-}

24. The central atom in the bromite ion BrO_2^- is surrounded by

a. two bonding and two unshared pairs of electrons.

b. three bonding and one unshared pair of electrons.

c. one bonding and three unshared pairs of electrons.

d. two double bonds and no unshared pairs of electrons.

e. four bonding and four lone pairs of electrons.

25. In the Lewis electron dot structure for hydrazine, N_2H_4, the total number of lone electron pairs around the two nitrogen atoms is

a. 0 b. 1 c. 2 d. 3 e. 4

26. In the Lewis structure for SF_4, the number of lone pairs of electrons around the central sulfur atom is

a. 0 b. 1 c. 2 d. 4 e. 5

27. Which one of the following species would have a Lewis structure most like that of SO_2?

a. CO_2 b. NO_2 c. NO_2^+ d. NO_2^- e. SiO_2

28. The number of valence electrons in the nitrite ion is

a. 16 b. 18 c. 22 d. 23 e. 24

29. The total number of valence electrons in the oxalate ion, $C_2O_4^{2-}$, is

a. 28 b. 30 c. 32 d. 34 e. 36

30. What is the average carbon-oxygen bond order in the formate ion?

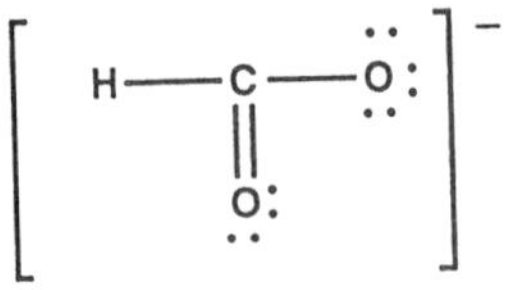

a. 0 b. 1 c. 1.5 d. 2 e. 2.5

31. Which statement is true regarding bond order, bond length, and bond energy.

a. As the bond order increases, the bond length increases.

b. As the bond order increases, the bond length decreases.

c. As the bond order increases, the bond energy decreases.

d. As the bond energy increases, the bond length increases.

e. As the bond energy increases, the bond order decreases.

32. Which of the following diatomic molecules has the greatest bond strength?

a. F_2 b. O_2 c. N_2 d. HF e. HCl

33. Given the bond dissociation energies below, calculate the standard molar enthalpy of formation of ClF_3.

$$Cl_2(g) + 3\ F_2(g) \rightarrow 2\ ClF_3(g)$$

Bond	Dissociation Energy (kJ/mol)
Cl-Cl	243
F-F	159
Cl-F	255

a. 210 kJ/mol b. 147 kJ/mol c. -33 kJ/mol d. -45 kJ/mol e. -405 kJ/mol

34. Given the bond dissociation energies below, calculate the standard molar enthalpy of formation of NF_3.

$$1/2\ N_2(g) + 3/2\ F_2(g) \rightarrow NF_3(g)$$

Bond	Dissociation Energy (kJ/mol)
N≡N	946
F-F	159
N-F	272

a. 833 kJ/mol b. 440. kJ/mol c. -104 kJ/mol d. -578 kJ/mol e. -618 kJ/mol

35. Xenon difluoride is prepared from elemental xenon and fluorine.

$$Xe(g) + F_2(g) \rightarrow XeF_2(g)$$

Calculate the enthalpy change, ΔH, for this reaction knowing that the bond dissociation energies are 131 kJ/mol for Xe-F and 159 kJ/mol for F-F.

a. 28 kJ b. -28 kJ c. -290. kJ d. -103 kJ e. 290 kJ

36. Which of the following best describes the energy change accompanying the process of breaking bonds in a molecule? (Ignore any subsequent reaction that may occur.)

a. Always endothermic.

b. Always exothermic.

c. The net energy change is always zero.

d. The change may be exothermic or endothermic depending on the physical state.

e. The change may be exothermic or endothermic depending on the substances involved.

37. Which compound contains a carbon-oxygen bond with a bond order of 2?

a. CO_2 b. CH_3OH c. CH_3OCH_3 d. CO e. C_2H_5OH

38. Which of the following groups of elements is arranged in order of increasing electronegativity?

a. Si < Al < Br < Cl b. Na < K < Ca < Ba c. P < S < O < F d. K < Rb < Cs < F

e. N < P < S < Cl

39. What is the formal charge on each atom in the following structure for the nitrite ion?

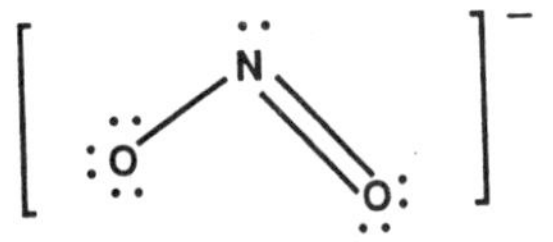

a. Nitrogen is 2-, oxygen on the left is 1-, oxygen on the right is 0.

b. Nitrogen is 0, oxygen on the left is 0, oxygen on the right is 1-.

c. Nitrogen is 0, oxygen on the left is 1-, oxygen on the right is 0.

d. Nitrogen is 3-, oxygen on the left is 1-, oxygen on the right is -2.

e. Nitrogen is 1+, oxygen on the left is 2-, oxygen on the right is 1-.

40. From a consideration of the Lewis structure shown below, what is the formal charge on sulfur in the molecule, SO_3?

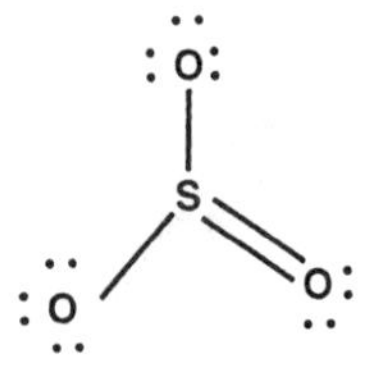

a. 0 b. 1+ c. 1- d. 2+ e. 2-

41. What are the oxidation numbers of sulfur and oxygen in the molecule SO_3?

a. Sulfur is +1 and oxygen is -1.

b. Sulfur is +6 and oxygen is -2.

c. Sulfur is +6 and oxygen is -6.

d. Sulfur is +3/2 and oxygen is -3.

e. Sulfur is -2 and oxygen is -2.

42. Using the VSEPR theory, predict the molecular shape of ClF_3.

a. triangular planar b. T-shaped c. linear d. tetrahedral e. square planar

43. Using the VSEPR theory, predict the molecular shape of SCl_2.

a. triangular planar b. T-shaped c. linear d. tetrahedral e. angular (bent)

44. Using the VSEPR theory, predict the molecular shape of the ion ICl_4-.

a. triangular planar b. angular (bent) c. octahedral d. tetrahedral e. square planar

45. Using the VSEPR theory, predict the molecular shape of NH_3.

a. triangular planar b. T-shaped c. triangular-pyramidal d. tetrahedral e. octahedral

46. Two ions which have a similar shape are

a. OH^- and SO_3^{2-} b. SO_3^{2-} and CO_3^{2-} c. PO_3^{3-} and CO_3^{2-} d. PO_3^{3-} and NO_3^-

e. SO_3^{2-} and PO_3^{3-}

47. Which of the following pairs of bonded atoms would be expected to have the longest bond length?

a. C - N b. C - S c. C - B d. C - F e. C - O

48. Which is the most polar bond?

a. O - F b. N - F c. C - F d. F - F e. Cl - F

49. What is the approximate C - O - H bond angle in CH_3OH?

a. 180° b. 120° c. 109.5° d. 90° e. 60°

50. What are the approximate bond angles of 1, 2, and 3 respectively?

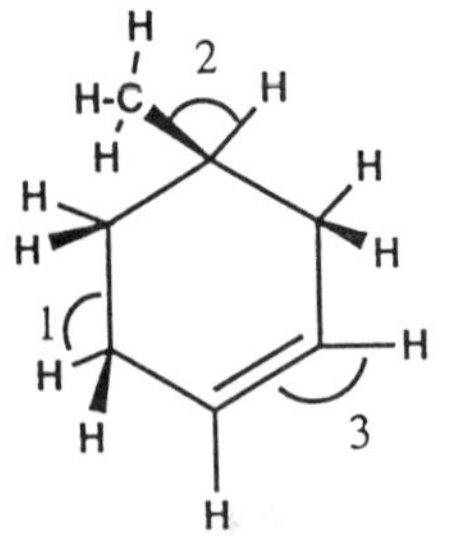

a. 120°, 120°, 180° b. 109.5°, 109.5°, 109.5° c. 109.5°, 120°, 120° d. 120°, 120°, 120°

e. 109.5°, 109.5°, 120°

Chapter 9: Answers:

1. c
2. b
3. e
4. a
5. c
6. d
7. c
8. b
9. a
10. e
11. a
12. c
13. c
14. a
15. d
16. e
17. d
18. b
19. c
20. b
21. a
22. c
23. e
24. a
25. c
26. b
27. d
28. b
29. d
30. c
31. b
32. c
33. e
34. c
35. d
36. a
37. a
38. c
39. c
40. d
41. b
42. b
43. e
44. e
45. c
46. e
47. b
48. c
49. c
50. e

Chapter 10
Bonding and Molecular Structure:
Orbital Hybridization and Molecular Orbitals

1. How many sigma (σ) bonds are in the following molecule?

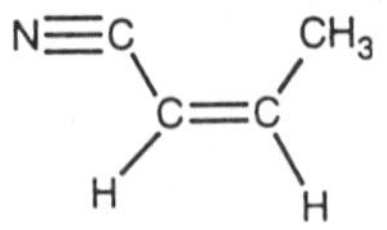

a. 3 b. 7 c. 8 d. 9 e. 12

2. How many pi (π) bonds are in the following molecule?

$CH_3C\equiv C\text{-}C\equiv N$

a. 2 b. 4 c. 6 d. 7 e. 10

3. How many sigma bonds (σ) are in the following molecule?

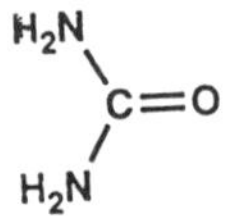

a. 3 b. 4 c. 6 d. 7 e. 8

4. In order to form a set of sp^2 hybrid orbitals, how many pure atomic orbitals of each type must be mixed?

a. one s and two p b. two s and two p c. two s and one p d. one s and three p e. two s and three p

5. Which of the following elements is most likely to display sp^3d hybridization?

a. oxygen b. nitrogen c. phosphorus d. carbon e. boron

6. Which of the following elements is most likely to display sp^3d^2 hybridization?

a. fluorine b. sulfur c. nitrogen d. carbon e. boron

7. What is the maximum number of hybrid orbitals that can be formed by oxygen?

a. two b. three c. four d. five e. six

8. How many sigma (σ) and pi (π) electron pairs are in a carbon dioxide molecule?

a. four σ and zero π b. three σ and two π c. two σ and two π d. two σ and four π

e. one σ and three π

9. How many sigma (σ) and pi (π) electron pairs are in a nitrogen molecule, N_2?

a. one σ and three π b. one σ and two π c. two σ and two π d. two σ and three π

e. one σ and three π

10. Assume that all the S-F bonds in SF_5^- are identical. The hybridization of the sulfur atomic orbitals which most likely account for this identity is

a. sp b. sp^2 c. sp^3 d. sp^3d e. sp^3d^2

11. What is the hybridization of the carbon atom in CS_2?

a. sp b. sp^2 c. sp^3 d. sp^3d e. sp^3d^2

12. What is the hybridization of the sulfur atom in SCl_4?

a. sp^3 b. sp^4 c. sp^3d d. sp^3d^2 e. sp^2d^2

13. What is the hybridization of the carbon atoms in CH_3OH and CO_2 respectively?

a. sp^3, sp^3 b. sp^3, sp^2 c. sp^2, sp^2 d. sp^2, sp^3 e. sp^3, sp

14. What is the hybridization of the oxygen atoms in CH_3OH and CO_2 respectively?

a. sp^3, sp^3 b. sp^3, sp^2 c. sp^2, sp^2 d. sp^2, sp^3 e. sp^3, sp

15. What is the hybridization of the nitrogen atoms in NH_3 and NH_4^+ respectively?

a. sp^3, sp^4 b. sp^3, sp^3 c. sp^2, sp^3 d. sp^2, sp^2 e. sp^3, sp

16. What is the hybridization of the sulfur atom in SO_3?

a. sp^2 b. sp^3 c. sp^4 d. sp^3d e. sp^3d^2

17. In the combustion of methane, CH_4, what change in hybridization (if any) occurs to the carbon atom?

a. sp^2 to sp^3 b. sp^2 to sp^3 c. sp^2 to sp^2 d. sp^3 to sp e. no change in hybridization occurs

18. In the addition of fluorine to xenon difluoride to form xenon tetrafluoride, what change in hybridization (if any) occurs to the xenon atom?

a. sp^3d^2 to sp^3d b. sp^2 to sp^3d c. sp^3 to sp^3d^2 d. sp^3d to sp^3d^2

e. no change in hybridization occurs

19. Which response contains all the characteristics listed that should apply to BF_3?

1. trigonal planar
2. one unshared pair of electrons on B
3. sp^2 hybridized boron atom
4. polar molecule
5. polar bonds

a. 2, 4, and 5 b. 1, 3, and 4 c. 1, 2, and 3 d. 1, 3, and 5 e. 3, 4, and 5

20. What hybrid orbitals of selenium are involved in the bonding in selenium dioxide?

a. sp b. sp^2 c. sp^3 d. sp^2d e. sp^3d

21. The nitrate ion is known to be planar with all the oxygen atoms equidistant from the central nitrogen atom. On the basis of these facts, which of the following conclusions is (are) **TRUE** concerning this ion?

1. It can be represented by three equivalent resonance structures.
2. The dipoles associated with each N-O bond are equal in magnitude.
3. The nitrogen atom is sp^2 hybridized.

a. 1 only b. 2 only c. 3 only d. 1 and 3 only e. 1, 2, and 3

22. What type of hybrid orbital set is used by the boron atom in the compound BF_3?

a. sp b. sp^2 c. sp^3 d. sp^3d e. sp^3d^2

23. What type of hybrid orbital set is used by the sulfur atom in the compound SF_6?

a. sp b. sp^2 c. sp^3 d. sp^3d e. sp^3d^2

24. What type of hybrid orbital set is used by the boron atom in the BCl_4^- ion?

a. sp b. sp^2 c. sp^3 d. sp^3d e. sp^3d^2

25. Which of the following statements is (are) **CORRECT** regarding π-bonding?

1. Pi (π) bonds do not occur unless the bonded atoms are already being joined by a sigma (σ) bond.
2. In order for a pi (π) bond to form, there must be an unhybridized p atomic orbital on the atom where the hybridization will occur.
3. The number of pi (π) bonds formed will equal the number of atomic orbitals on the hybridized atom.

a. 1 only b. 2 only c. 3 only d. 1 and 2 only e. 1, 2, and 3

26. What hybrid orbital set is used by the central carbon atom in the allene molecule?

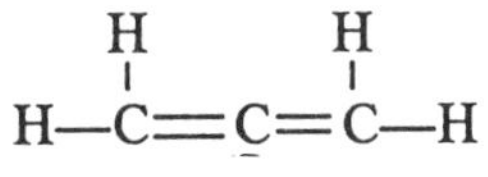

$$H-\overset{\overset{\textstyle H}{|}}{C}=\underset{}{C}=\overset{\overset{\textstyle H}{|}}{C}-H$$

a. sp   b. sp^2 c. sp^3 d. sp^3d e. sp^3d^2

27. What hybrid orbital set is used by the terminal carbon atoms in the following molecule?

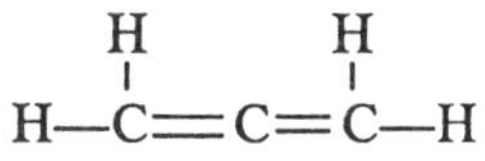

$$H-\overset{\overset{\textstyle H}{|}}{C}=C=\overset{\overset{\textstyle H}{|}}{C}-H$$

a. sp   b. sp^2 c. sp^3 d. sp^3d e. sp^3d^2

28. Methylbenzoate, called "oil of Niobe," is used in perfumes.

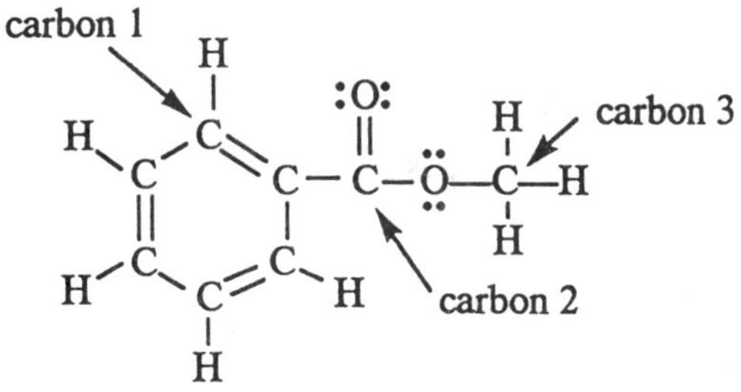

The hybridizations of carbons 1, 2, and 3 in the figure above are described respectively as

a. sp, sp, sp^2 b. sp, sp^2, sp^3 c. sp^2, sp^2, sp^3 d. sp^2, sp^2, sp^2 e. sp, sp, sp^3

29. The number of sigma bonds in SCN^- is

a. 1   b. 2   c. 3   d. 4   e. 5

30. The number of π-bonds in hydrazine, H_2NNH_2 is

a. 0   b. 1   c. 2   d. 3   e. 4

31. All of the following species contain two π-bonds **EXCEPT**

a. SCN^- b. CO   c. H_2CCO d. OCS   e. NO^-

32. A molecular orbital that decreases the electron probability between the nuclei is said to be _____.

a. antibonding.   b. nonbonding.   c. bonding.   d. hybridized.   e. diamagnetic.

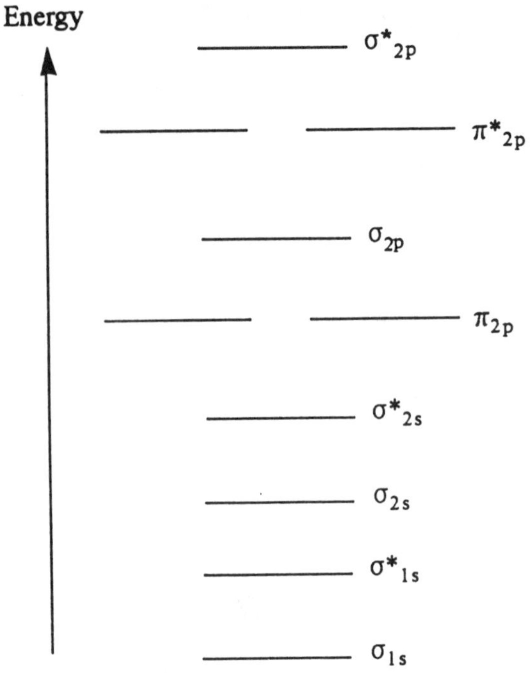

33. All of the following statements regarding molecule orbital theory are correct **EXCEPT**

a. The number of molecular orbitals formed is equal to the number of atomic orbitals contributed by the atoms involved.

b. Electrons are assigned to orbitals of successively higher energy.

c. The Pauli Principle is obeyed.

d. A bonding molecular orbital is higher in energy than the parent atomic orbital from which it was formed.

e. Hund's rule is observed.

34. In a diatomic molecule, when two atomic orbitals of the same type combine to form molecular orbitals, which of the following statements best describes the energy of the resulting pair of molecular orbitals?

a. Both molecular orbitals will be higher in energy than the component atomic orbitals.

b. Both molecular orbitals will be lower in energy than the component atomic orbitals.

c. One molecular orbital will be higher in energy and one will be lower in energy than the component atomic orbitals.

d. Both molecular orbitals will be identical in energy to the component atomic orbitals.

e. One molecular orbital will be antibonding and the other will be nonbonding.

35. According to molecular orbital theory, which of the following species is unlikely to exist?

a. H_2 b. H_2^+ c. He_2^+ d. He_2 e. H_2^-

36. According to the molecular orbital theory, which of the following correctly lists the following oxygen species in terms of **increasing** bond order.

a. $O_2^+ < O_2^- < O_2$ b. $O_2^+ < O_2 < O_2^-$ c. $O_2^- < O_2^+ < O_2$ d. $O_2^- < O_2 < O_2^+$
e. $O_2 < O^{2+} < O_2^-$

37. Consider the diatomic molecules of the second period Li_2, Be_2, and C_2. Which is (are) unlikely to exist?

a. Li_2 b. Li_2 and Be_2 c. Be_2 d. C_2 e. Be_2 and C_2

38. Consider the diatomic molecules of the second period Li_2, Be_2, B_2, and C_2. The two molecules which have the same bond order are

a. Li_2 and Be_2 b. Li_2 and B_2 c. Li_2 and C_2 d. Be_2 and B_2 e. Be_2 and C_2

39. Consider the diatomic molecules of the second period C_2, N_2, O_2, and F_2. The two molecules which have the same bond order are

a. C_2 and N_2 b. N_2 and O_2 c. O_2 and F_2 d. N_2 and O_2 e. C_2 and O_2

40. The species below having the longest bond (if any) is

a. N_2^+ b. N_2 c. N_2^- d. N_2^{2-} e. They are all the same length.

41. The species below having the shortest bond (if any) is

a. N_2^+ b. N_2 c. N_2^- d. N_2^{2-} e. They are all the same length.

42. The following molecular orbital energy level diagram is appropriate for which one of the listed particles?

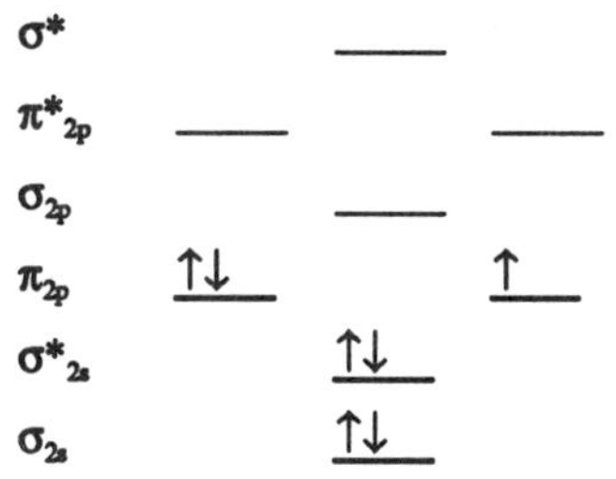

a. B_2^+ b. B_2^- c. N_2^+ d. N_2^- e. N_2

43. The following molecular orbital energy level diagram is appropriate for which one of the listed particles?

σ^*_{2p}		___	
π^*_{2p}	↑		↑
σ_{2p}		↑↓	
π_{2p}	↑↓		↑↓
σ^*_{2s}		↑↓	
σ_{2s}		↑↓	

a. O_2^{2-} b. B_2 c. F_2^- d. N_2^{2-} e. N_2

44. The molecular orbital configuration of B_2 is

a. [core electrons] $(\sigma_{2s})^2 (\sigma^*_{2s})^2 (\pi_{2p})^4 (\sigma_{2p})^2$

b. [core electrons] $(\sigma_{2s})^2 (\sigma^*_{2s})^2 (\pi_{2p})^4 (\sigma_{2p})^1 (\pi^*_{2p})^1$

c. [core electrons] $(\sigma_{2s})^2 (\sigma^*_{2s})^2 (\pi_{2p})^2 (\sigma_{2p})^1$

d. [core electrons] $(\sigma_{2s})^2 (\sigma^*_{2s})^2 (\pi_{2p})^1$

e. [core electrons] $(\sigma_{2s})^2 (\sigma^*_{2s})^2 (\pi_{2p})^2$

45. Of the species, N_2, N_2^+, N_2^-, and N_2^{2-}, how many are paramagnetic?

a. four b. three c. two d. one e. zero

46. The elements called __________ are characterized by the fact that the valence band is only partly filled.

a. metals b. semiconductors c. nonmetals d. allotropes e. metalloids

47. Which substance has cations bonded together by mobile electrons?

a. Ag(s) b. S_8(s) c. Br_2(ℓ) d. KBr(s) e. $MgCl_2$(s)

48. In order to create an *n-type* semiconductor, a silicon crystal could be doped with

a. Ga b. Ge c. As d. He e. None of these.

49. In order to create an *p-type* semiconductor, a silicon crystal could be doped with

a. Ga b. Ge c. As d. He e. None of these.

50. How do the number of electrons in the valence band of a metal compare with an insulator?

a. completely full for a metal; partially filled for an insulator

b. completely full for a metal; completely full for an insulator

c. partially filled for a metal; partially filled for an insulator

d. partially filled for a metal; completely full for an insulator

e. completely filled for a metal; totally empty for an insulator

Chapter 10: Answers:

1. d	26. a
2. b	27. b
3. d	28. c
4. a	29. b
5. c	30. a
6. b	31. e
7. c	32. a
8. c	33. d
9. b	34. c
10. e	35. d
11. a	36. d
12. c	37. c
13. e	38. b
14. b	39. e
15. b	40. d
16. a	41. b
17. d	42. b
18. d	43. d
19. d	44. e
20. b	45. b
21. e	46. a
22. b	47. a
23. e	48. c
24. c	49. a
25. d	50. d

Chapter 11
Bonding and Molecular Structure:
Carbon -- More Than Just Another Element

1. Which of the following formulas represents an alkane?

a. C_2H_4 b. C_3H_7 c. C_4H_{10} d. C_5H_8 e. C_6H_6

2. Which of the following formulas could represent a cycloalkene?

a. C_2H_4 b. C_3H_7 c. C_4H_{10} d. C_5H_8 e. C_6H_6

3. Which of the following formulas could represent an alkyne?

a. C_2H_4 b. C_3H_7 c. C_3H_6 d. C_5H_8 e. C_6H_{14}

4. The chemical formula of acetylene is

a. CH_4 b. C_2H_2 c. C_6H_6 d. C_2H_4 e. CH_3CO_2H

5. How many of the following compounds are unsaturated?

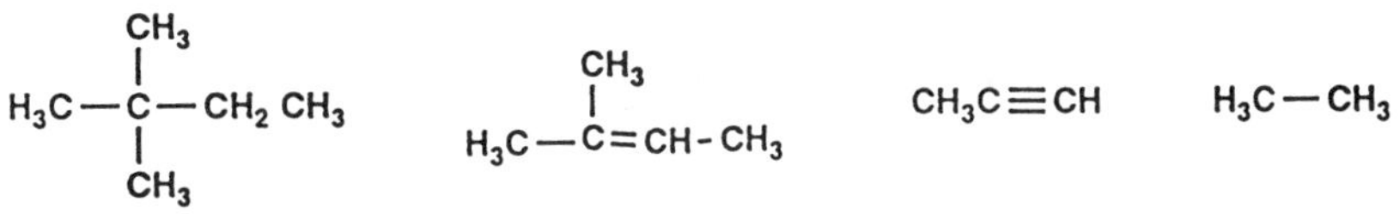

a. zero b. 1 c. 2 d. 3 e. 4

6. The formula for the following compound is 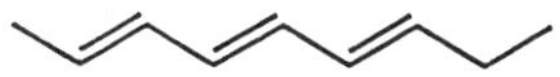

a. C_9H_{14} b. C_9H_{12} c. C_7H_8 d. C_7H_{12} e. C_8H_{12}

7. The formula for the following compound is

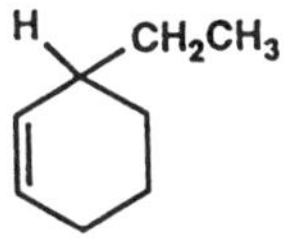

a. C_6H_6 b. C_6H_{10} c. C_6H_{12} d. C_8H_{14} e. C_8H_{16}

8. The formula for the following compound is

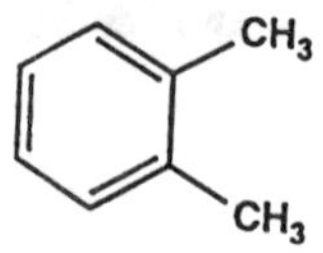

a. C_8H_6 b. C_6H_6 c. C_6H_{10} d. C_8H_{10} e. C_8H_{12}

9. Which of the following formulas represents pentane?

a. C_4H_8 b. C_4H_{10} c. C_5H_{10} d. C_5H_{12} e. C_6H_{12}

10. How many hydrogen atoms are needed to complete the following hydrocarbon structure?

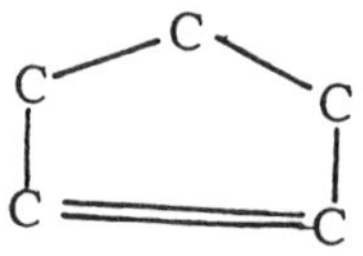

a. 4 b. 6 c. 8 d. 10 e. 12

11. A straight-chain alkene has six carbon atoms. Its molecular formula is

a. C_6H_6 b. C_6H_8 c. C_6H_{10} d. C_6H_{12} e. C_6H_{14}

12. What are the bonds present in butrene, C_4H_8?

a. 9 σ bonds and 3 π bonds b. 10 σ bonds and 2 π bonds c. 11 σ bonds and 1 π bond

d. 12 σ bonds and 1 π bond e. 12 σ bonds

13. The bond angles in propene are

a. 109.5° b. 120° c. 109.5° and 120° d. 109.5°, 120°, and 180° e. 90°, 115°, and 120°

14. The number of carbon atoms in the molecule, 2, 3, 4, trimethyl-5-heptene is

a. 9 b. 10 c. 11 d. 12 e. 13

15. Which statement best describes organic compounds?

a. Most are soluble in water.

b. Most are soluble in water after treatment with NaOH(aq).

c. Most have only covalent bonding.

d. Most have high melting points.

e. Most are gases.

16. The name of the following compound is

$$H_3C-\underset{\underset{CH_3}{|}}{\overset{\overset{CH_2CH_2CH_3}{|}}{C}}-CH_2\ CH_3$$

   a. 2,2-diethylpropane.   b. 2-methyldibutane.   c. 2-methyloctane.   d. 2-methyl-2-propylbutane.
   e. 3,3-dimethylhexane.

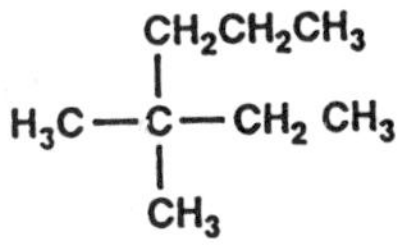

17. The name of the following compound is

$$H_3C-\underset{\underset{CH_3}{|}}{\overset{\overset{CH_3}{|}}{C}}-CH_2\ CH_3$$

   a. 2,2,3-trimethylpropane.   b. 2,2-dimethylpropane.   c. 2,2-dimethylbutane.
   d. 2,2-dimethylhexane.   e. 3,3-dimethylhexane.

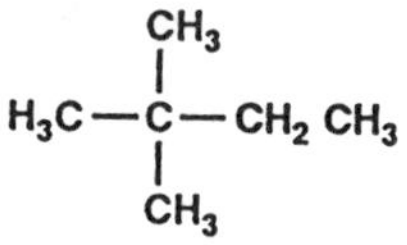

18. The balanced equation for the complete combustion of 2-butene is
   a. $C_4H_6 + 6\ O_2 \rightarrow 4CO_2 + 3H_2O$
   b. $C_4H_{10} + 4\ O_2 \rightarrow 4CO_2 + 5H_2$
   c. $C_4H_8 + 4\ O_2 \rightarrow 4C + 4H_2O$
   d. $C_4H_8 + 6\ O_2 \rightarrow 4\ CO_2 + 4H_2O$
   e. $C_4H_{10} + 13/2\ O_2 \rightarrow 4CO_2 + 5H_2O$

19. How many of the following compounds are aromatic molecules?

   a. zero   b. 1   c. 2   d. 3   e. 4

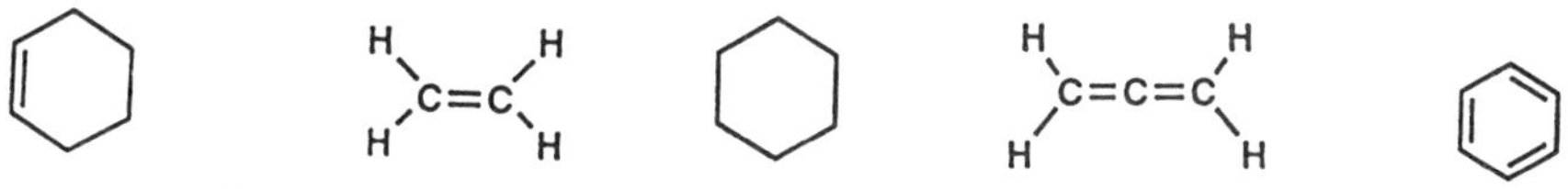

20. The common name of the following compound is

   a. o-dimethylbenzene.   b. p-dimethylbenzene.   c. m-dimethylbenzene.   d. 2,3-dimethylbenzene.
   e. toluene.

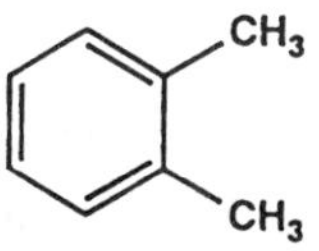

21. Which of the following are polar molecules?

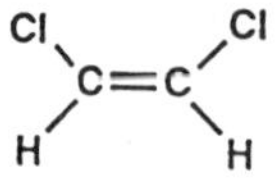

#1

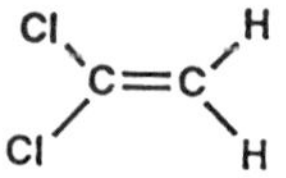

#2

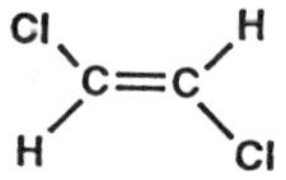

#3

a. 1 only b. 2 only c. 3 only d. 1 and 2 only e. 2 and 3 only

22. What is the product of the addition of Br_2 to $H_2C{=}CH_2$?

a. 1,1-dibromoethylene b. 1,1-dibromoethane c. 1,2-dibromoethylene d. 1,2-dibromoethane

e. 1,2-dibromocyclopropane

23. What is the product of the addition of H_2O to $H_2C{=}CH_2$?

a. ethanol b. methanol c. 1-propanol d. 2-propyl alcohol e. 1,2-ethandiol

24. Which of the following alcohols is the poisonous "wood alcohol"?

a. glycerol b. ethanol c. methanol d. 1-propanol e. 2-propanol

25. Which of the following molecules is a secondary alcohol?

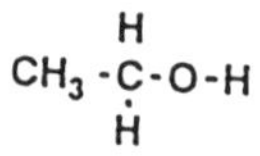

a.

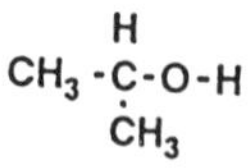

b.

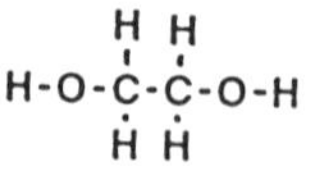

c.

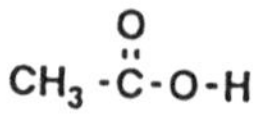

d.

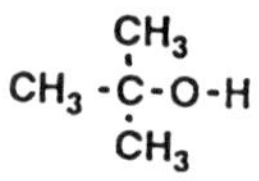

e.

26. An isomer of dimethyl ether CH_3-O-CH_3 is

a. $CH_3CO_2CH_3$ b. CH_3CH_2-O-CH_2CH_3 c. $HOCH_2CH_2OH$ d. CH_3CO_2H e. CH_3CH_2OH

27. When an alcohol is dehydrated, the product could be an

a. alkane. b. alkene. c. aldehyde. d. acid. e. ester.

28. When a primary alcohol is oxidized, the initial product is an

a. alkane. b. alkene. c. aldehyde. d. alkyne. e. isomer.

29. Lactic acid, found in milk, has the formula

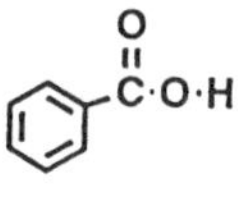

a.

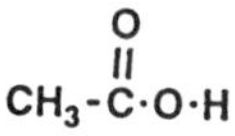

b.

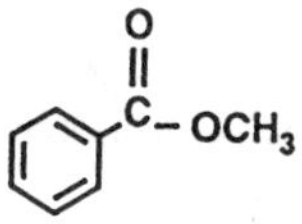

c.

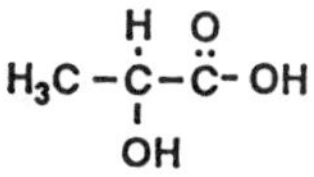

d.

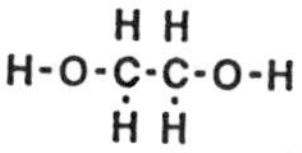

e.

30. The functional group $-\overset{O}{\overset{\|}{C}}-O-H$ is characteristic of

a. acids. b. esters. c. alcohols. d. ethers. e. ketones.

31. The functional group - O - H is characteristic of

a. acids. b. bases. c. alcohols. d. ethers. e. aldehydes.

32. Which functional group does **not** contain an oxygen atom?

a. ester b. amide c. alkene d. aldehyde e. alcohol

33. The C = O linkage occurs in molecules with the following functional groups **EXCEPT**

d. ketones. b. aldehydes. c. alkenes. d. carboxylic acids. e. esters.

34. The C = O linkage occurs in molecules with the following functional groups **EXCEPT**

a. amines. b. amides. c. esters. d. carboxylic acids. e. ketones.

35. Which of the following functional groups contains a C = O linkage?

a. alcohols b. ethers c. amines d. amides e. alkynes

36. From a consideration of the following:

1 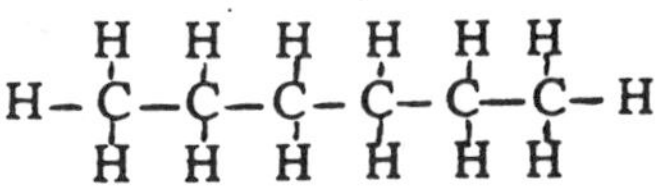

2 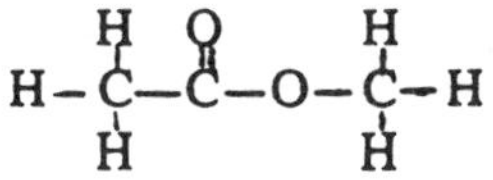

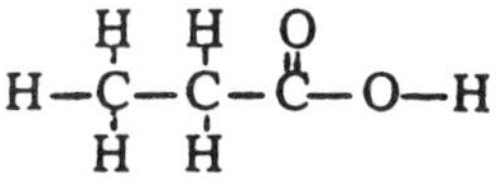

3

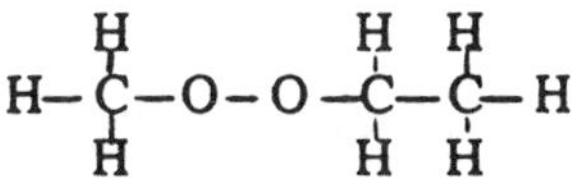

4

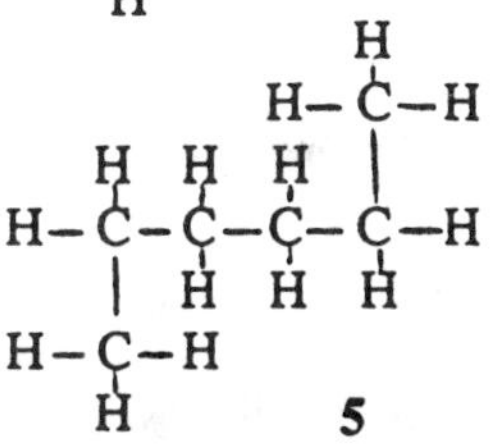

5

Which pair of compounds are isomers?

a. 1 and 2 b. 2 and 3 c. 2 and 4 d. 1 and 5 e. 3 and 4

37. The condensation reaction of ethanol and formic acid will yield as products water and an

a. ester. b. ether. c. alkene. d. aldehyde. e. anhydride.

38. The number of different structural isomers of dichlorobenzene is

a. 1 b. 2 c. 3 d. 4 e. 6

39. Which of the following pairs are isomers?

a. $CH_3CH_2CH_2OH$ and CH_3CH_2CHO

b. $CH_3CH_2OCH_3$ and $CH_3CH_2CH_2OH$

c. CH_3COOH and CH_3CHO

d. $CH_3CH_2CH(CH_3)CH_3$ and $CH_3CH(CH_3)CH_2CH_3$

e. $CH_3CH = CH_2$ and $CH_2 = CHCH_3$

40. Which group of compounds includes an aldehyde, an acid, and an alcohol (in any order)?

a. HCO_2H, $CH_3CO_2CH_3$, CH_3CH_2OH

b. H_2CO, CH_3CH_2OH, $CH_3CO_2CH_3$

c. CH_3CO_2H, CH_3OH, $CH_3CH_2OCH_3$

d. H_2CO, CH_3CO_2H, CH_3CHO

e. H_2CO, CH_3CO_2H, CH_3CH_2OH

41. Which of the following compounds is an ester?

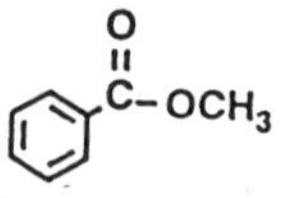

a.

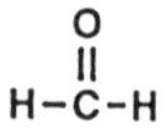

b.

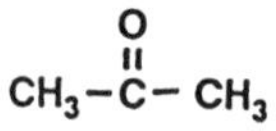

c.

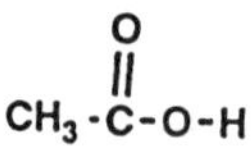

d.

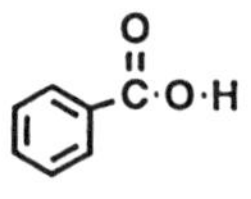

e.

42. When methanol is heated with benzoic acid, in the presence of acid, the product is

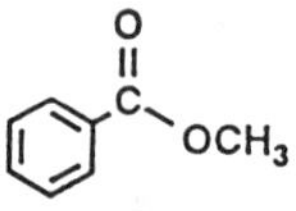

a.

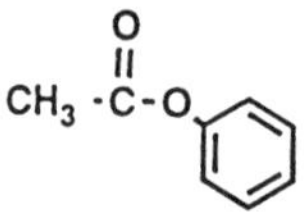

b.

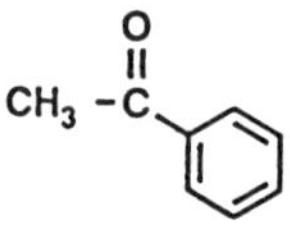

c.

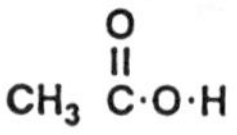

d.

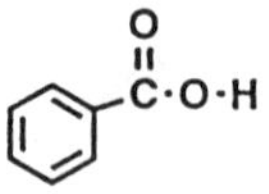

e.

43. The molecule

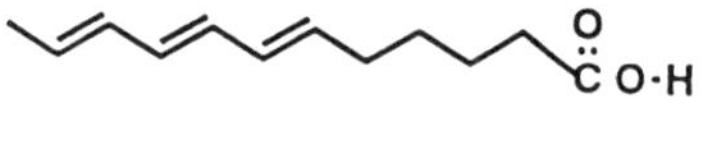

is classified as

a. an aromatic acid. b. a fatty acid. c. a soap. d. an aldehyde. e. a polyester.

44. The monomer of polystyrene is

a. $C_6H_5CH = CH_2$ b. $C_6H_5CH_3$ c. $C_6H_5CH_2CH_3$ d. $H_2C = CH_2$ e. $H_2C = CHCH = CH_2$

45. Polyethylene

a. contains equal numbers of cis and trans double bonds.

b. contains no double bonds.

c. reacts with methanol to form polyester.

d. cannot form branched chains.

e. is an example of a condensation polymer.

46. Low density polyethylene

a. packs together easily with long linear molecules.

b. has a branched structure.

c. is used to make tough rigid materials such as milk cartons.

d. cannot be recycled.

e. contains long chains of alternating double bonds.

47. Which of the following statements is (are) **TRUE** about polymers?

1. They are macromolecules.
2. They have simple repeating units.
3. They are all made by condensation reactions.

a. 1 only b. 2 only c. 3 only d. 1 and 2 only e. 1, 2, and 3

48. A copolymer of styrene and butadiene is used to make what common product?

a. coffee cups b. plastic wrap for foods c. synthetic rubber d. paint e. nylon

49. The condensation polymer formed by a diacid and a dialcohol is classified as a

a. composite. b. soap. c. polyurethane. d. polyamide. e. polyester.

50. Nylon-6,6 can be produced from the condensation polymerization of a diacid with six carbon atoms and a diamine with six carbon atoms. Nylon is classified as a

a. polyamine. b. polyurethane. c. polyalcohol. d. polyamide. e. polyester.

Chapter 11: Answers:

1. c	26. e
2. d	27. b
3. d	28. c
4. b	29. d
5. c	30. a
6. a	31. c
7. d	32. c
8. d	33. c
9. d	34. a
10. c	35. d
11. d	36. b
12. c	37. a
13. c	38. c
14. b	39. b
15. c	40. e
16. e	41. a
17. c	42. a
18. d	43. b
19. b	44. a
20. a	45. b
21. d	46. b
22. d	47. d
23. a	48. c
24. c	49. e
25. b	50. d

Chapter 12
Gases and Their Properties

1. Which of the following represents the largest gas pressure?

   a. 5.0 torr b. 5.0 mm Hg c. 5.0 atm d. 5.0 kPa e. 5.0 bar

2. Which of the following represents the smallest gas pressure?

   a. 1.8 lb/in^2 b. 1.8 torr c. 1.8 atm d. 1.8 kPa e. 1.8 bar

3. Rank 355 torr, 0.524 atm, and 0.513 bar in increasing order of pressure.

   a. 355 torr < 0.524 atm < 0.513 bar b. 0.524 atm < 355 torr < 0.513 bar
   c. 0.513 bar < 355 torr < 0.524 atm d. 355 torr < 0.513 bar < 0.524 atm
   e. 0.513 bar < 0.524 atm < 355 torr

4. What are standard temperature and pressure conditions for gases?

   a. 0°C and 0 torr b. 0 K and 760 torr c. -273°C and 1 atm d. 0°C and 760 torr
   e. 0°C and 1 torr

5. If the volume of a confined gas is doubled while the temperature remains constant, what change (if any) would be observed in the pressure?

   a. It would be half as large. b. It would double. c. It would be four times as large.
   d. It would be 1/4 as large. e. It would remain the same.

6. A given mass of gas in a rigid container is heated from 100°C to 500°C. Which of the following responses best describes what will happen to the pressure of the gas?

   a. The pressure will decrease by a factor of five.
   b. The pressure will increase by a factor of five.
   c. The pressure will increase by a factor of about two.
   d. The pressure will increase by a factor of about eight.
   e. The pressure will increase by a factor of about twenty-five.

7. Which of the following has the most molecules?

   a. 1.00 L of CH_4 at 0°C and 1.00 atm b. 1.00 L of N_2 at 0°C and 1.00 atm
   c. 1.00 L of O_2 at 20°C and 1.00 atm d. 1.00 L of CO_2 at 50°C and 1.25 atm
   e. 1.00 L of CO at 0°C and 1.25 atm

8. Avogadro stated that equal volumes of gases under the same conditions of temperature and pressure have equal

a. numbers of molecules. b. numbers of grams. c. molar masses. d. atoms. e. speeds.

9. A given mass of gas occupies the volume of 5.00 L at 65°C and 480 mm Hg. Which mathematical expression represents the volume of the gas at 630 mm Hg and 85°C?

a. $5.00 \times \frac{65}{85} \times \frac{480}{630}$ b. $5.00 \times \frac{338}{358} \times \frac{480}{630}$ c. $5.00 \times \frac{358}{338} \times \frac{480}{630}$ d. $5.00 \times \frac{358}{338} \times \frac{630}{480}$

e. $5.00 \times \frac{338}{358} \times \frac{630}{480}$

10. A gas occupies a volume of 1.50 L at 400 mm Hg and 100°C. Which mathematical expression gives the correct volume at 700 mm Hg and 200°C?

a. $1.50 \times \frac{400}{700} \times \frac{373}{473}$ b. $1.50 \times \frac{400}{700} \times \frac{473}{373}$ c. $1.50 \times \frac{700}{400} \times \frac{373}{473}$ d. $1.50 \times \frac{700}{400} \times \frac{473}{373}$

e. $1.50 \times \frac{400}{700} \times \frac{200}{100}$

11. What volume of CH_4 at 0°C and 1.00 atm contains the same number of molecules as 0.50 L of N_2 measured at 27°C and 1.50 atm?

a. 0.37 L b. 0.46 L c. 0.68 L d. 0.50 L e. 0.82 L

12. What volume of SO_2 at 25°C and 1.50 atm contains the same number of molecules as 2.00 L of Cl_2 measured at 0°C and 1.00 atm?

a. 0.68 L b. 1.22 L c. 1.45 L d. 1.83 L e. 2.18 L

13. If 5.42 L of air measured at 735 torr and 23°C is heated to 35°C in the same container, what is the new pressure?

a. 13.4 torr b. 333 torr c. 414 torr d. 520 torr e. 765 torr

14. If 3.0 L of helium at 20.0°C is allowed to expand to 4.4 L, with the pressure remaining the same, what is the new temperature?

a. 702°C b. 430°C c. 157°C d. -30.0°C e. -55°C

15. A 4.50 L flask of Ar at 23°C and 734 torr is heated to 55°C. What is the new pressure?

a. 366 torr b. 935 torr c. 1.25 torr d. 1.06 atm e. 2.58 atm

16. An air compressor reduced a sample of helium originally at 25°C and 740 torr to 6.75 liters at 42.0 atm and 85°C. What was the original volume of the helium?

a. 85.6 liters b. 35.9 liters c. 242 liters d. 319 liters e. 350 liters

17. At what temperature will 41.6 grams N_2 exerts a pressure of 815 torr in a 20.0 L cylinder?

a. 134 K b. 176 K c. 238 K d, 337 K e. 400 K

18. What volume will a mixture of 0.200 mole N_2 and 0.500 mole He occupy at 0.944 atm and 15.0°C?

a. 0.913 liters b. 5.00 liters c. 12.5 liters d. 15.7 liters e. 17.5 liters

19. When 7.00 grams of helium and 14.0 grams of argon were mixed in a flask, the pressure was measured as 712 torr. What is the partial pressure of the helium?

a. 593 torr b. 356 torr c. 833 torr d. 1070 torr e. 1420 torr

20. What pressure (in atmospheres) is exerted by 82.5 grams of CH_4 in a 75.0 liter container at 35.0°C?

a. 0.197 atm b. 0.339 atm c. 1.73 atm d. 2.57 atm e. 27.8 atm

21. At STP, what is the volume of a mixture of gases containing 1.0 mol Ar, 1.5 mol He, and 2.5 mol Kr?

a. 5.0 liters b. 17.5 liters c. 22.4 liters d. 112 liters e. 345 liters

22. A mixture of the gases neon and krypton is in a 2.00 liter container. The partial pressure of the neon is 0.40 atm and the partial pressure of the krypton is 1.20 atm. What is the mol fraction of neon?

a. 0.20 b. 0.25 c. 0.33 d. 0.60 e. 0.80

23. When 0.34 moles of He are mixed with 0.51 moles of Ar in a flask, the total pressure in the flask is found to be 5.0 atm. What is the partial pressure of Ar in this flask?

a. 0.85 atm b. 1.5 atm c. 2.0 atm d. 3.0 atm e. 5.0 atm

24. If 20.0 mL of $SO_2(g)$ and 20.0 mL of $Cl_2(g)$ react according to the equation below, what is the total volume of all gases after the reaction when they are at the same temperature and pressure?

$$SO_2(g) + 2Cl_2(g) \rightarrow OSCl_2(g) + Cl_2O(g)$$

a. 20.0 mL b. 26.6 mL c. 30.0 mL d. 40.0 mL e. 66.6 mL

25. Which of the following gases has the greatest density at 0°C and 1 atm?

a. N_2 b. 0_2 c. F_2 d. Ne e. CO

26. Calculate the density of SO_3 gas at 35°C and 715 torr.

a. 0.0285 g/L b. 1.43 g/L c. 2.15 g/L d. 2.98 g/L e. 3.57 g/L

27. What is the density of CH_4 at 200°C and 0.115 atm?

a. 0.0475 g/L b. 0.0716 g/L c. 0.542 g/L d. 0.870 g/L e. 2.09 g/L

28. What is the density of SO_2 at 83°C and 1.30 atm?

a. 3.73 g/L b. 2.85 g/L c. 1.72 g/L d. 1.10 g/L e. 0.582 g/L

29. At what temperature will the density of Ar be 2.66 g/L when the pressure is 1.32 atm?

a. -230°C b. -74°C c. -32°C d. 17°C e. 43°C

30. What is the molar mass of a gas which has a density of 1.30 g/L measured at 27°C and 0.400 atm?

a. 38.0 g/mol b. 48.0 g/mol c. 61.5 g/mol d. 80.0 g/mol e. 97.5 g/mol

31. What is the molar mass of a gas which has a density of 1.83 g/L measured at 27°C and 0.538 atm?

a. 25.0 g/mol b. 38.0 g/mol c. 45.8 g/mol d. 75.4 g/mol e. 83.7 g/mol

32. What is the chemical formula of a gas if it has a pressure of 1.40 atm and a density of 1.82 g/L at 27°C?

a. CO_2 b. CO c. CH_4 d. O_2 e. N_2

33. What is the chemical formula of a gas if it has a pressure of 680 torr and a density of 1.60 g/L at 27°C?

a. C_2H_6 b. CO_2 c. NO d. F_2 e. CF_4

34. At standard conditions, it was found that 1.12 L of a gas weighed 2.78 g. Its molecular weight is

a. 2.78 g/mol b. 27.8 g/mol c. 55.6 g/mol d. 111 g/mol

e. Not enough information is given to solve this problem.

35. The volume of a certain gas sample is 235 mL when collected over water at a temperature of 25°C and a pressure of 698 mmHg. What will be the volume of this gas sample when measured dry at standard pressure? The vapor pressure of water at this temperature is 23.8 mmHg.

a. 197 mL b. 208 mL c. 223 mL d. 265 mL e. 275 mL

36. A sample of oxygen gas collected by displacement of water at 40.0 °C and a pressure of 691 mmHg has a volume of 534 mL. Calculate the volume that this sample of oxygen will occupy when dry under standard conditions of temperature and pressure. The vapor pressure of water is 55.3 mmHg at 40.0°C.

a. 591 mL b. 531 mL c. 443 mL d. 404 mL e. 390. mL

37. The empirical formula of a certain hydrocarbon is CH_2. When 0.125 moles of this hydrocarbon is completely combusted with excess oxygen, it is observed that 8.40 liters of CO_2 gas are produced at STP. What is the molecular formula of the unknown hydrocarbon?

a. CH_2 b. C_2H_4 c. C_2H_3 d. C_3H_6 e. C_4H_8

38. The empirical formula of a certain hydrocarbon is CH_2. When 0.125 moles of this hydrocarbon is completely combusted with excess oxygen, it is observed that 11.2 L of H_2O gas is produced at STP. What is the molecular formula of the unknown hydrocarbon?

a. C_2H_4 b. C_2H_3 c. C_3H_6 d. C_4H_8 e. C_6H_{12}

39. Complete combustion of a 0.20 mol sample of a hydrocarbon, C_xH_y, gives 0.80 mol of CO_2 and 1.0 mol of H_2O. The molecular formula of the original hydrocarbon is

a. C_3H_8 b. C_4H_5 c. C_4H_8 d. C_4H_{10} e. C_8H_{20}

40. What volume of O_2, collected at 22.0°C and 728 mmHg would be produced by the decomposition of 8.15 g $KClO_3$?

$$2KClO_3(s) \rightarrow 2KCl(s) + 3\,O_2(g)$$

a. 1.12 L b. 1.48 L c. 1.68 L d. 2.23 L e. 2.52 L

41. What volume of O_2, measured at 25°C and 733 torr, is needed for the complete combustion of 42.0 grams of ethylene, C_2H_4?

a. 38.0 L b. 44.9 L c. 62.7 L d. 114 L e. 341 L

42. What volume of O_2, measured at 27.0°C and 0.750 atm, is needed for the complete combustion of 24.0 grams of propylene, C_3H_8?

a. 112 L b. 89.5 L c. 60.9 L d. 53.6 L e. 17.8 L

43. Ammonia gas is synthesized according to the balanced equation

$$N_2(g) + 3H_2(g) \rightarrow 2NH_3(g)$$

If 15.0 liters of nitrogen are reacted with an excess of hydrogen, how many liters of ammonia could be produced? Assume all gas volumes are measured at the same temperature and pressure.

a. 5.00 L b. 10.0 L c. 15.0 L d. 20.0 L e. 30.0 L

44. Carbon dioxide gas diffuses through a porous barrier at a rate of 0.20 mL/minute. If an unknown gas diffuses through the same barrier at a rate of 0.313 mL/minute, what is the molar mass of the unknown gas?

a. 28 g/mole b. 35 g/mole c. 39 g/mole d. 68 g/mole e. 84 g/mole

45. How long will it take for 10.0 milliliters of helium gas to effuse through a certain porous barrier if it has been previously observed that it required 84 minutes for 10.0 milliliters of chlorine gas to effuse through that same barrier?

a. 4.8 min b. 12 min c. 20 min d. 28 min e. 350 min

46. Which of the following statements is **true**?

a. All particles moving with the same velocity have the same kinetic energy.
b. All particles at the same temperature have the same kinetic energy.
c. All particles having the same kinetic energy have the same mass.
d. As the kinetic energy of a particle is halved, its velocity is also halved.
e. As the velocity of a particle is doubled, the kinetic energy decreases by a factor of four.

47. Which of the following gases effuses at the highest rate?

a. N_2 b. O_2 c. F_2 d. Ne e. CO

48. At a particular temperature, which of the following molecules has an average velocity closest to that of ethylene, C_2H_4, at the same temperature?

a. N_2 b. CO_2 c. NO_2 d. O_2 e. CH_4

49. Non-ideal behavior for a gas is most likely to be observed under conditions of

a. standard temperature and pressure.
b. low temperature and high pressure.
c. low temperature and low pressure.
d. high temperature and high pressure.
e. high temperature and low pressure.

50. For a given sample of gas molecules, the average kinetic energy depends only on the value of the

a. pressure. b. temperature. c. volume. d. moles. e. molar mass.

Chapter 12: Answers:

1. c	26. d
2. b	27. a
3. d	28. b
4. d	29. c
5. a	30. d
6. c	31. e
7. e	32. d
8. a	33. b
9. c	34. c
10. b	35. b
11. c	36. e
12. c	37. d
13. e	38. d
14. c	39. d
15. d	40. e
16. c	41. d
17. b	42. b
18. e	43. e
19. a	44. a
20. c	45. c
21. d	46. b
22. b	47. d
23. d	48. a
24. c	49. b
25. c	50. b

Chapter 13
Bonding and Molecular Structure:
Intermolecular Forces, Liquids, and Solids

1. Which one of the following substances would exhibit dipole-dipole intermolecular forces?

a. CCl_4 b. Cl_2 c. N_2 d. NCl_3 e. CH_4

2. The forces that exist between noble gas atoms in the liquid and solid state are

a. dipole/dipole. b. ion/dipole. c. induced dipole/induced dipole. d. hydrogen bonds. e. ion/ion.

3. At room temperature, which of the following compounds has the strongest interparticle forces?

a. CO_2 b. H_2O c. NaCl d. CH_3CH_3 e. CH_3Cl

4. Which of the following would be expected to have the highest boiling point?

a. Ar b. He c. Kr d. Ne e. Xe

5. Of the gases, Ne, N_2, O_2, CH_4, and SiH_4, which one would you expect to be easiest to liquefy?

a. Ne b. N_2 c. O_2 d. CH_4 e. SiH_4

6. Which of the following would probably have the lowest boiling point?

a. CH_4 b. SiH_4 c. PH_3 d. AsH_3 e. NH_3

7. Which of the following molecules would exhibit hydrogen bonding in the liquid state?

a. CH_4 b. H_2 c. NH_3 d. CH_3F e. H_2S

8. Which of the following would be expected to have the highest melting point?

a. Cl_2 b. NaCl c. $MgCl_2$ d. CCl_4 e. CH_4

9. Which of the following would be expected to have the highest melting point?

a. LiBr b. LiF c. LiCl d. NaCl e. NaI

10. Which of the following would be expected to have the lowest melting point?

a. KI b. $CaCl_2$ c. NaBr d. KBr e. MgF_2

11. Which of the following substances would be expected to have the highest melting point?

a. CaO b. Br_2 c. CO_2 d. HCl e. H_2O

12. Which of the following solids is held in the solid state primarily by induced dipole-induced dipole forces?

a. ice b. NaCl c. solid NH_3 at low temperature d. I_2 e. Na

13. Which of the following molecules is most polar?

a. Cl_2 b. HCl c. NO d. IBr e. H_2

14. Rank the compounds NH_3, CH_4, and SiH_4 in order of increasing boiling point.

a. $NH_3 < CH_4 < SiH_4$ b. $CH_4 < NH_3 < SiH_4$ c. $NH_3 < SiH_4 < CH_4$ d. $CH_4 < SiH_4 < NH_3$

e. $SiH_4 < NH_3 < CH_4$

15. If KBr, C_2H_5OH, C_2H_6, and He are arranged in order of increasing boiling point, the list is

a. He $< C_2H_5OH < C_2H_6 <$ KBr b. KBr $< C_2H_6 < C_2H_5OH <$ He c. He $< C_2H_6 < C_2H_5OH <$ KBr

d. KBr $< C_2H_6 <$ He $< C_2H_5OH$ e. $C_2H_5OH < C_2H_6 <$ He $<$ KBr

16. Liquids with high surface tension tend to have

a. strong intermolecular forces. b. strong adhesive forces. c. strong intramolecular forces.

d. weak cohesive forces. e. no cohesive forces.

17. When KCl dissolves in water, what types of intermolecular bonds are formed?

a. ion-ion b. hydrogen bonds c. ion-dipole d. ion-ion forces and H-bonds e. dipole-dipole

18. When NaBr dissolves in water, what types of intermolecular forces must be broken?

a. ion-ion forces. b. H-bonds c. ion-dipole forces d. ion-ion forces and H-bonds

e. dipole-dipole

19. The equation which represents the number of atoms in a face-centered cubic unit cell is

a. 8(1/8) + 4(1/2) b. 4(1/4) + 4 c. 6(1/4) + 6(1/2) d. 8(1/8) + 4(1/4) + 2(1/2)

e. 8(1/8) + 6(1/2)

20. An atom that is shared equally between eight cubic unit cells is called

a. an edge atom. b. a face atom. c. a corner atom. d. a diagonal atom. e. a central atom.

21. The volume of a body-centered cubic unit cell (bcc) of tungsten is 3.31×10^{-23} cm^3. The density of tungsten is

a. 36.8 g/cm^3 b. 18.4 g/cm^3 c. 9.22 g/cm^3 d. 5.55 g/cm^3 e. 2.44 g/cm^3

22. The mass of a unit cell of calcium is 2.66 x 10^{-22} g. Calculate the number of atoms of calcium in the unit cell.

a. 1 b. 2 c. 4 d. 6 e. 8

23. Chromium crystallizes in a body-centered cubic unit cell. In a unit cell of chromium each atom is surrounded by how many "nearest neighbors"?

a. 2 b. 4 c. 6 d. 8 e. 12

24. Gold crystallizes as a face-centered cubic lattice with a unit cell edge of 407.86 pm.

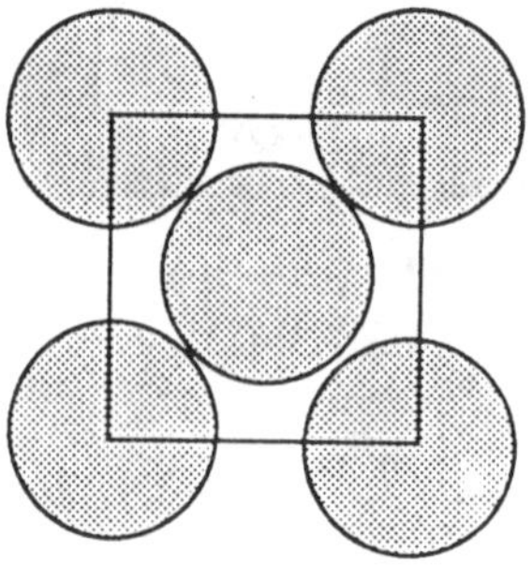

Calculate the radius of a gold atom.

a. 102.0 pm b. 134.0 pm c. 144.2 pm d. 203.9 pm e. 407.9 pm

25. The figure at the right represents a unit cell where copper atoms are on the faces and gold atoms are at the corners. The empirical formula of the alloy is

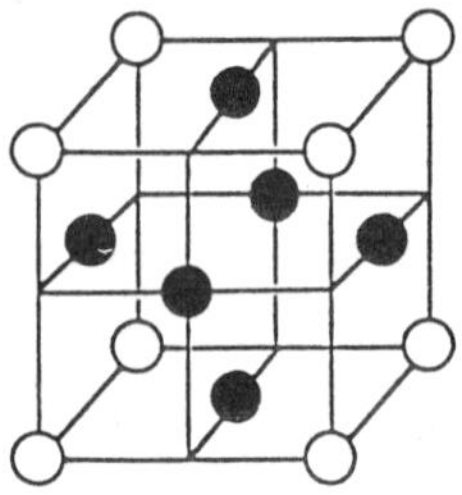

a. $CuAu_2$ b. Cu_3Au c. Cu_6Au_8
d. $CuAu_6$ e. Cu_6Au

26. In the unit cell at the right element X is within the cell and element Y is at the corners. The formula of the compound is

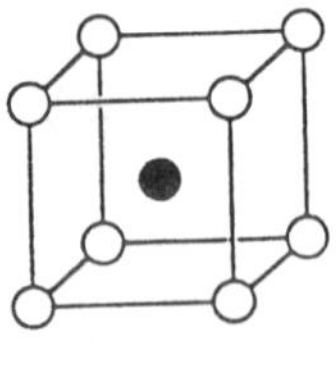

a. XY b. XY_2 c. XY_4 d. XY_6 e. XY_8

27. A metal fluoride crystallizes such that the fluoride ions occupy cubic lattice positions at the corners and on the faces while the 4 metal atoms occupy positions within the body of the unit cells. The formula of the metal fluoride is

a. MF b. MF_2 c. MF_3 d. M_4F_{14} e. MF_8

Questions 28 - 32 pertain to lead (atomic mass of 207.2 g/mol) which crystallizes in a face-centered cubic arrangements. Lead has an atomic radius of 1.75×10^{-8} cm.

28. Calculate the length of the edge of the unit cell in centimeters.

a. 1.24×10^{-8} cm b. 1.75×10^{-8} cm c. 2.48×10^{-8} cm d. 3.50×10^{-8} e. 4.95×10^{-8} cm

29. What is the volume of the unit cell in cm^3?

a. 1.21×10^{-22} cm^3 b. 1.53×10^{-22} cm^3 c. 1.91×10^{-22} cm^3 d. 4.28×10^{-22} cm^3
e. 5.36×10^{-22} cm^3

30. How many lead atoms are there per unit cell?

a. 1 b. 2 c. 4 d. 6 e. 8

31. What is the mass of the unit cell in grams?

a. 3.44×10^{-22} g/cell b. 6.88×10^{-22} g/cell c. 1.03×10^{-21} g/cell d. 1.38×10^{-21} g/cell
e. 2.75×10^{-22} g/cell

32. What is the density of lead in g/cm^3?

a. 9.85 g/cm^3 b. 11.4 g/cm^3 c. 13.2 g/cm^3 d. 14.7 g/cm^3 e. 19.7 g/cm^3

Questions 33 - 38 pertain to gold (atomic mass 197.0 g/mol) which crystallizes in a face-centered cubic arrangement with the atoms touching along the face diagonal. A gold atom has a radius of 1.44×10^{-8} cm.

33. Calculate the length of the edge of the unit cell in centimeters.

a. 1.44×10^{-8} cm b. 2.88×10^{-8} cm c. 3.56×10^{-8} cm d. 4.07×10^{-8} cm e. 4.38×10^{-8} cm

34. What is the volume of the unit cell in cm^3?

a. 4.58×10^{-23} cm^3 b. 4.72×10^{-23} cm^3 c. 5.02×10^{-23} cm^3 d. 6.23×10^{-23} cm^3
e. 6.76×10^{-23} cm^3

35. How many gold atoms are there per unit cell?

a. 1 b. 2 c. 4 d. 8 e. 12

36. What is the mass of one gold atom?

a. 3.27×10^{-22} g b. 4.25×10^{-22} g c. 197.0 g d. 4.25×10^{23} g e. 3.27×10^{23} g

37. What is the mass of the unit cell in grams?

a. 3.27×10^{-22} g b. 4.25×10^{-22} g c. 6.54×10^{-22} g d. 8.50×10^{-22} e. 1.31×10^{-21} g

38. What is the density of gold in g/cm^3?

a. 12.8 g/cm^3 b. 16.9 g/cm^3 c. 19.4 g/cm^3 d. 20.4 g/cm^3 e. 23.8 g/cm^3

39. Which of the following properties of water can be attributed to hydrogen bonding?

1. High surface tension
2. High heat of vaporization
3. Low vapor pressure

a. 1 only b. 2 only c. 3 only d. 1 and 2 only e. 1, 2, and 3

40. Methanol, CH_3OH, (molar mass 32.04 g/mol) has a heat of vaporization of 39.2 kJ/mol and a density of 0.7914 g/mL. How much energy is needed to vaporize 350. mL of methanol?

a. 1.08×10^4 kJ b. 8.86×10^2 kJ c. 428 kJ d. 652 kJ e. 339 kJ

41. Which of the following processes (if any) is exothermic?

a. solid → gas b. solid → liquid c. liquid → solid d. liquid → gas e. none of these

42. A chemist sets up two beakers of distilled water under the same room conditions in a laboratory. One beaker is boiling vigorously, the other beaker is boiling gently. Which of the following statements is true?

a. The temperature of the vigorously boiling water is higher.

b. The temperature of the gently boiling water is higher.

c. The temperature of the water in both beakers is the same.

d. The boiling points of the water in the two beakers must be different.

e. The temperature in the vigorously boiling water is not uniform.

43. As the pressure on a pure substance is increased at constant temperature, a phase transition (if any) which is **not** observed would be

a. gas → solid b. liquid → solid c. solid → gas d. gas → liquid

e. all of these could be observed

44. The greatest change in energy for a substance is seen with which process?

a. vaporization b. condensation c. fusion d. sublimation e. melting

45. Which of the following exhibit **NO CHANGE IN TEMPERATURE** when 27 joules of heat energy is removed?

a. 100 g $H_2O(s)$ at -10°C b. 100 g $H_2O(\ell)$ at 0°C c. 100 g $H_2O(\ell)$ at 75°C

d. 100 g $H_2O(g)$ at 110°C e. 100 g $H_2O(s)$ at 0°C

46. From a consideration of the phase diagram below, the changes in a substance upon going from points A to B to C to D correspond to

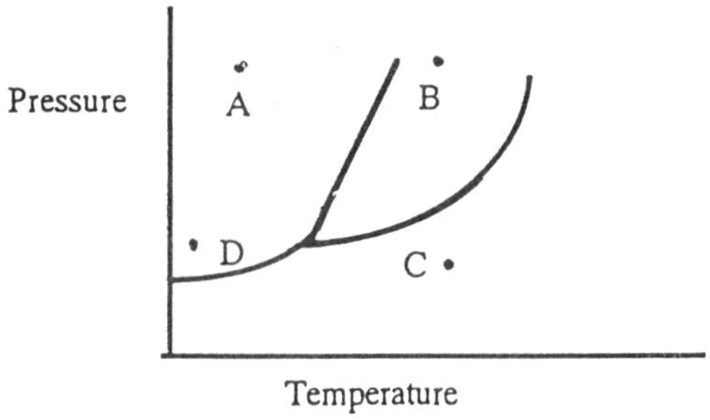

a. melting, vaporization, deposition

b. vaporization, freezing, sublimation

c. sublimation, freezing, melting

d. freezing, sublimation, vaporization

e. melting, sublimation, deposition

47. Which of the following is an example of a network solid?

a. SiO_2 b. MgO c. P_4 d. NaCl e. I_2

48. The sublimation of solid carbon dioxide, dry ice, is an example of

a. evaporation. b. a physical property. c. a chemical property. d. a chemical change.
e. a physical change.

49. The heat of sublimation of a compound equals

a. heat of fusion plus heat of vaporization.

b. heat of ionization plus heat of crystallization.

c. heat of vaporization minus heat of fusion.

d. heat of vaporization plus heat of crystallization.

e. heat of crystallization minus heat of vaporization.

50. The critical point of carbon tetrachloride is 283°C and 45 atm pressure. Liquid carbon tetrachloride has a vapor pressure of 10.0 atm at 178°C. Which of the following statements must be true?

a. The normal boiling of CCl_4 must be greater than 178°C.

b. Liquid CCl_4 can exist at temperatures greater than 283°C if the pressure is greater than 45 atm.

c. The triple point must be less than 178°C.

d. Liquid and solid can only be in equilibrium at one temperature -- the freezing point.

e. Vapor and liquid can only be in equilibrium at one temperature -- the normal boiling point.

Chapter 13: Answers:

1. d	26. a
2. c	27. a
3. c	28. e
4. e	29. a
5. e	30. c
6. a	31. d
7. c	32. b
8. c	33. d
9. b	34. e
10. a	35. c
11. a	36. a
12. d	37. e
13. b	38. c
14. d	39. e
15. c	40. e
16. a	41. c
17. c	42. c
18. d	43. c
19. a	44. d
20. c	45. b
21. b	46. a
22. c	47. a
23. d	48. e
24. c	49. a
25. b	50. c

Chapter 14
Solutions and Their Behavior

1. What is the molality of 7.80% by weight glucose ($C_6H_{12}O_6$ molar mass = 180.16 g/mol) solution?

a. 0.470 m　b. 0.845 m　c. 0.0432 m　d. 0.0454 m　e. 0.0844 m

2. What is the molality of a 5.45% by weight Na_2SO_4 (molar mass = 142.06 g/mol) solution?

a. 0.0383 m　b. 0.0818 m　c. 0.406 m　d. 7.74 m　e. 8.18 m

3. The mol fraction of NH_4Cl in a solution is 0.0311. What is its molality? (The molar mass of water is 18.016 g/mol.)

a. 1.78 m　b. 1.66 m　c. 0.969 m　d. 0.562 m　e. 0.0983 m

4. What is the weight percent $ZnCl_2$ of a 1.12 m $ZnCl_2$ (molar mass = 136.29 g/mol) solution?

a. 1.52%　b. 7.55%　c. 13.2%　d. 18.0%　e. 24.6%

5. What is the mol fraction Na_2SO_4 in a solution which is 11.5% by weight Na_2SO_4 (molar mass Na_2SO_4 = 142.06 g/mol and H_2O = 18.016 g/mol)?

a. 0.0810　b. 0.0914　c. 0.0745　d. 0.0173　e. 0.0162

6. If the mol fraction NaCl in a solution is 0.0175, what is the weight percent NaCl (molar mass NaCl = 58.44 g/mol and H_2O = 18.016 g/mol)?

a. 5.46%　b. 5.77%　c. 10.2%　d. 11.5%　e. 17.7%

7. What is the mol fraction $NaNO_3$ in a solution which is 2.15 m?

a. 0.0180　b. 0.0268　c. 0.0373　d. 0.09387　e. 0.0785

8. What is the weight percent $FeCl_3$ (molar mass = 162.22 g/mol) in a solution which is 1.84 m?

a. 14.0%　b. 16.2%　c. 29.8%　d. 25.2%　e. 23.0%

9. A 1.34 M $NiCl_2$ (molar mass = 129.6 g/mol) solution has a density of 1.12 g/cm^3. What is the weight percent $NiCl_2$ of the solution?

a. 1.73%　b. 8.64%　c. 15.5%　d. 25.4%　e. 29.8%

10. A 1.34 M $NiCl_2$ (molar mass = 129.6 g/mol) solution has a density of 1.12 g/cm^3. What is the molality of the solution?

a. 0.913 m　b. 1.42 m　c. 1.55 m　d. 2.55 m　e. 3.13 m

11. A 1.26 M $Cu(NO_3)_2$ (molar mass = 187.56 g/mol) solution has a density of 1.19 g/cm^3. What is the weight percent $Cu(NO_3)_2$ of the solution?

a. 1.88% b. 2.36% c. 10.5% d. 14.3% e. 19.9%

12. What is the molarity of a 25.0% HCl solution if the density is 1.08 g/cm^3?

a. 2.70 M b. 2.96 M c. 5.49 M d. 7.41 M e. 9.11 M

13. Hydrobromic acid (molar mass = 80.9 g/mol) is commercially available in a 34.0 mass percent solution which has a density of 1.31 g/cm^3. What is the molarity of the commercially available hydrobromic acid?

a. 2.75 M b. 4.45 M c. 5.50 M d. 9.35 M e. 10.2 M

14. A solution of hydrogen peroxide is 30.0% H_2O_2 by weight and has a density of 1.11 g/cm^3. The molarity of the solution is

a. 13.3 M b. 9.78 M c. 8.82 M d. 7.94 M e. 0.980 M

15. The maximum contamination level of arsenic ion in a water system is 0.050 parts per million. If the arsenic is present as $AsCl_3$, how many grams of arsenic chloride could be present in a system that contains 8.2×10^5 liters?

a. 0.55 g b. 7.3 g c. 41 g d. 62 g e. 98 g

16. A solution has a magnesium ion concentration of 1.2×10^3 ppm. If the magnesium ion is dissolved as $MgCl_2$, how many grams of $MgCl_2$ are in each liter of solution?

a. 1.1 g b. 1.3 g c. 4.7 g d. 4.9 g e. 5.8 g

17. A student prepared a solution containing 0.30 mol solute and 1.00 mole solvent. The mole fraction of solvent is

a. 1.30 b. 1.00 c. 0.77 d. 0.30 e. 0.23

18. What is the molality of ions in a solution containing 23.4 grams of $Mg(NO_3)_2$ (molar mass = 148.3 g/mol) in 964 grams of water?

a. 0.158 m b. 0.164 m c. 0.491 m d. 0.628 m e. 0.823 m

19. Which measure of concentration is most appropriate for the calculation of the vapor pressure of a solution?

a. mol fraction b. molarity c. molality d. weight % e. ppm

20. If 0.1 gram of a compound with a molar mass of 10,000 is dissolved in 50.0 grams of water, what colligative property is most appropriate to measure in order to determine its molar mass?

a. freezing point depression b. vapor pressure lowering c. molarity d. osmotic pressure e. molality

21. A chemist knows the empirical formula of a new compound but not the molecular formula. What must be determined experimentally so that the molecular formula can be determined.

a. density b. viscosity c. % composition d. melting point e. molar mass

22. The amount of solvent (grams or moles) is known for each of the following solution concentrations **EXCEPT**

a. molarity. b. molality. c. mass %. d. mole fraction. e. ppm.

23. A volumetric flask is necessary for the preparation of which one of the following concentration measurements?

a. molality b. X c. mass % d. molarity e. ppm

24. Which of the following substances is more soluble in hexane (C_6H_{14}) than in water?

a. $CaCl_2$ b. CH_3OH c. NH_3 d. $C_8H_{17}Cl$ e. N_2H_4

25. If the pressure of a gas over a liquid increases, the amount of gas dissolved in the liquid will

a. increase. b. decrease. c. remain the same. d. have a higher vapor pressure.

e. depends on the polarity of the gas.

26. Which of the following gases do you expect to have the highest Henry's Law gas constant for water at 25°C?

a. CO_2 b. HCN c. N_2 d. O_2 e. Ne

27. Which of the following solvents should Na^+ ion have the highest solvation energy in?

a. $H_2O(\ell)$ b. $CS_2(\ell)$ c. $CCl_4(\ell)$ d. $CH_3CH_2CH_2CH_2CH_2CH_3(\ell)$ e. $CH_3OH(\ell)$

28. Which of the following changes in the property of a salt would increase its heat of hydration?

a. a decrease (weakening) in lattice energy

b. a decrease in hydration energy of the cation

c. a decrease in hydration energy of the anion

d. an increase (strengthening) in lattice energy

e. an increase in the melting point of the salt

29. What is the primary energetic factor in the lack of miscibility between $CCl_4(\ell)$ and water?

a. the strength of intermolecular forces between CCl_4 molecules

b. the strength of intermolecular forces between H_2O molecules

c. the charge on the C atom in CCl_4

d. the difference between the molecular weights of the molecules

e. the electronegativity difference between carbon and chlorine

30. The heat of solution of NH_4NO_3 is +25.7 kJ/mol. We have a beaker that contains 100. mL of a saturated solution of NH_4NO_3 at 25°C. The beaker has 52 grams of undissolved solid on the bottom. Which of the following changes will cause some of the dissolved NH_4NO_3 to precipitate?

a. remove 10. grams of the solid from the beaker

b. heat the solution by 10. degrees

c. cool the solution by 10. degrees

d. add 10. grams of additional NH_4NO_3 solid

e. add 10. grams of NaCl solid

31. Which of the following solutions would have the lowest vapor pressure?

a. 1 m glucose ($C_6H_{12}O_6$) b. 1 m $MgCl_2$ c. 1 m $NaNO_3$ d. 1 m NaBr e. pure H_2O

32. Which of the following solutions would have the highest vapor pressure?

a. pure H_2O b. 1 m glucose ($C_6H_{12}O_6$) c. 1 m $NaNO_3$ d. 1 m $MgCl_2$ e. 1 m $(NH_4)_2SO_4$

33. Which of the following would have a boiling point closest to that of 1 m NaCl?

a. pure H_2O b. 1 m sucrose ($C_{12}H_{22}O_{11}$) c. 1 m $MgCl_2$ d. 0.5 m CH_3OH e. 1 m NH_4NO_3

34. Which of the following solutions would have a freezing point closest to that of a 1 molal solution of $CaCl_2$?

a. 1 m $CaSO_4$ b. 1 m KBr c. 1 m Na_2SO_4 d. 0.5 m $SnCl_4$ e. pure H_2O

35. Which of the following would have the highest freezing point?

a. 1 m glucose ($C_6H_{12}O_6$) b. 1 m $MgCl_2$ c. 1 m $NaNO_3$ d. 1 m $(NH_4)_2SO_4$ e. pure H_2O

36. Which of the following would have the lowest freezing point?

a. pure H_2O b. 1 m urea (CON_2H_4) c. 1 m KCl d. 1 m $NaNO_3$ e. 1 m Na_2SO_4

37. You need a solution that is 0.15 m in ions. How many grams of $MgCl_2$ (molar mass = 95.2 g/mol) must you dissolve in 400. g of water? (Assume total dissociation of the ionic salt.)

a. 0.060 g b. 1.9 g c. 5.7 g d. 7.6 g e. 17 g

38. We dissolve 50.0 grams of acetic acid, H_3CCOOH (molar mass = 60.05 g/mol) in 500. grams of water at 25°C. What is the vapor pressure of water for this solution? Vapor pressure for water at 25°C = 23.8 mmHg.

a. 0.690 mmHg b. 12.8 mmHg c. 23.1 mmHg d. 27.2 mmHg e. 32.6 mmHg

39. What is the freezing point of a solution containing 4.134 grams naphthalene (molar mass = 128.2) dissolved in 30.0 grams paradichlorobenzene? The freezing point of pure paradichlorobenzene is 53.0°C and the freezing point depression constant K_{fp} is -7.10°C/m.

a. 52.0°C b. 48.7°C c. 45.4°C d. 17.6°C e. 7.63°C

40. What is the molar mass of a nonelectrolyte if 5.48 grams dissolved in 35.0 grams of benzene freezes at -1.39°C? The freezing point of pure benzene is 5.50°C and the freezing point depression constant K_{fp} is -5.12°C/m.

a. 116 g/mol b. 189 g/mol c. 245 g/mol d. 412 g/mol e. 609 g/mol

41. A solution is prepared by dissolving 0.500 g of non-dissociating solute in 12.0 g of cyclohexane. The freezing point depression of the solution is 8.94°C. The K_{fp} for cyclohexane is -20.0°C/m. Calculate the molar mass of the solute.

a. 93.2 g/mol b. 112 g/mol c. 128 g/mol d. 182 g/mol e. 205 g/mol

42. What is the molar mass of a compound if 5.96 grams is dissolved in 25.0 grams of chloroform solvent to form a solution which has a boiling point elevation of 4.80°C. The boiling point constant of chloroform is +3.63°C/m.

a. 112 g/mol b. 132 g/mol c. 180 g/mol d. 342 g/mol e. 451 g/mol

43. What is the molar mass of a compound if 4.28 grams is dissolved in 25.0 grams of chloroform solvent to form a solution which has a boiling point elevation of 2.30°C. The boiling point constant of chloroform K_{bp} is +363°C/m.

a. 35.4 g/mol b. 67.5 g/mol c. 135 g/mol d. 168 g/mol e. 270 g/mol

44. The osmotic pressure of blood is 7.65 atm at 37°C. How many grams of glucose ($C_6H_{12}O_6$, molar mass = 180.2 g/mol) are needed to prepare 1.00 liter of a solution for intravenous injection that has the same osmotic pressure as blood?

a. 3.00 g b. 4.44 g c. 25.4 g d. 45.3 g e. 54.1 g

45. What is the osmotic pressure of a 0.100 M aqueous solution of urea at 25°C?

a. 244 atm b. 24.4 atm c. 3.82 atm d. 2.44 atm e. 0.244 atm

46. All of the following are examples of colloidal dispersions EXCEPT

a. a saturated solution of NaCl. b. milk. c. pudding. d. steel. e. mayonnaise.

47. Concentrated salt solutions have boiling points lower than those calculated using the equation $\Delta T_{bp} = K_{bp} \cdot m$. Which of the following is a reasonable explanation of this observation?

a. Positive ions repel each other more at high concentration.

b. Ions of opposite charge will tend to stay paired instead of breaking up.

c. The water molecules will have a greater attraction for each other.

d. Concentrated solutions really have small particles of non-dissolved salt, thus lowering the molality.

e. The difference between the crystal lattice energy and the heat of hydration must be taken into consideration.

48. Which of the following is a real life example of osmotic pressure?

a. distilling alcohol b. salting roads for de-icing c. salting meats for preservation
d. the use of anti-freeze in cars e. increasing the CO_2 pressure in soft drinks

49. When Solution A, 0.10 M NaCl, and Solution B, 0.20 M NaCl, are separated by a semipermeable membrane, what occurs during osmosis?

a. Solvent molecules move from B into A.

b. Na^+ and Cl^- ions move from B into A.

c. The vapor pressure of A increases.

d. The molarity of A increases.

e. No change is observed.

50. The hydrophobic portion of a soap molecule is

a. the long carbon chain. b. the carboxylic acid. c. the glycerol. d. the anion of the acid.
e. the short carbon chain.

Chapter 14: Answers:

1. a	26. b
2. c	27. a
3. a	28. a
4. c	29. b
5. e	30. c
6. a	31. b
7. c	32. a
8. e	33. e
9. c	34. c
10. b	35. e
11. e	36. e
12. d	37. b
13. c	38. c
14. b	39. c
15. e	40. a
16. c	41. a
17. c	42. c
18. c	43. e
19. a	44. e
20. d	45. d
21. e	46. a
22. a	47. b
23. d	48. c
24. d	49. d
25. a	50. a

Chapter 15
Principles of Reactivity: Chemical Kinetics

1. Reaction rates usually

a. increase with time. b. decrease with time. c. multiply with time. d. double with time.

e. remain constant.

2. For the gas phase reaction , $3H_2 + N_2 \rightarrow 2NH_3$, how does the rate of disappearance of H_2 compare to the rate of production of NH_3?

a. The initial rates are equal.

b. The rate of disappearance of H_2 is 1/2 the rate of appearance of NH_3.

c. The rate of disappearance of H_2 is 3/2 the rate of appearance of NH_3.

d. The rate of disappearance of H_2 is 2/3 the rate of appearance of NH_3.

e. The rate of disappearance of H_2 is 1/3 the rate of appearance of NH_3.

3. Consider the reaction

$$S_2O_8^{2-} + 3\,I^- \rightarrow 2SO_4^{2-} + I_3^-$$

which one of the following rate expressions would give the same value as the rate of disappearance of $S_2O_8^{2-}$?

a. rate = $-3\ \Delta[I^-] / \Delta t$ b. rate = $-1/3\ (\Delta[I^-]) / \Delta t$ c. rate = $-2(\Delta[SO_4^{2-}])\ \Delta t$

d. rate = $-\Delta[I_3^-] / \Delta t$ e. rate = $-1/2\ (\Delta[SO_4^{2-}]) / \Delta t$

4. For the reaction

$$Cl_2 + 2NO \rightarrow 2NOCl$$

which one of the following rate expressions would give the same value for the rate as the change in molarity of Cl_2?

a. $-2\ \Delta[NO] / \Delta t$ b. $+1/2\ (\Delta[NOCl]) / \Delta t$ c. $-2\ (\Delta[NOCl] / \Delta t)$ d. $-1/2\ (\Delta[NOCl] / \Delta t)$

e. $+1/2\ (\Delta[NO] / \Delta t)$

5. For the reaction

$$6CH_2O + 4NH_3 \rightarrow (CH_2)_6N_4 + 6H_2O$$

the rate is expressed as $\frac{1}{6}\frac{\Delta[H_2O]}{\Delta t}$. An equivalent would be

a. $-\frac{\Delta[(CH_2)_6N_4]}{\Delta t}$ b. $6\frac{\Delta[CH_2O]}{\Delta t}$ c. $-6\frac{\Delta[CH_2O]}{\Delta t}$ d. $-\frac{1}{4}\frac{\Delta[NH_3]}{\Delta t}$ e. $-\frac{1}{6}\frac{\Delta[H_2O]}{\Delta t}$

6. Given the initial rate data for the reaction A + B $\rightarrow$ C, determine the rate expression for the reaction

[A], M	[B], M	Δ[C]/Δt (initial) M/s
0.10	0.20	5.00
0.20	0.20	10.0
0.10	0.15	2.81

a. $\Delta[C]/\Delta t = 1250[A][B]^2$ b. $\Delta[C]/\Delta t = 250[A][B]$ c. $\Delta[C]/\Delta t = 250[A]^2$

d. $\Delta[C]/\Delta t = 50.0[A]$ e. $\Delta C/\Delta t = 5.0[A][B]$

7. Given the initial rate data for the reaction A + B $\rightarrow$ C, determine the rate expression for the reaction

[A], M	[B], M	Δ[C]/Δt (initial) M/s
0.13	0.28	1.11
0.13	0.15	0.596
0.24	0.15	2.03

a. $\Delta[C]/\Delta t = 109[A][B]^2$ b. $\Delta[C]/\Delta t = 30.5[A][B]$ c. $\Delta[C]/\Delta t = 235[A]^2[B]$

d. $\Delta[C]/\Delta t = 8.53[A]$ e. $\Delta[C]/\Delta t = 235[A][B]$

8. The reaction

$$CH_3CHO(g) \rightarrow CH_4(g) + CO(g)$$

proceeds via the rate expression $\Delta[CO]/\Delta t = [CH_3CHO]^{3/2}$. What is the overall order of the reaction?

a. zero-order b. first-order c. second-order d. third-order e. three-halves-order

9. For a reaction, the rate law is rate = $k[A]^1[B]^0[C]^1$. What are the units for k where the time unit is seconds (s)?

a. (mol/L·s) b. L/mol·s c. $L^2/mol^2 \cdot s$ d. $mol^2/L^2 \cdot s$ e. $mol^3/L^3 \cdot s$

10. The rate law for the chemical reaction

$$5Br^- + BrO_3^- + 6H^+ \rightarrow 3Br_2 + 3H_2O$$

has been determined experimentally to be

$$\frac{-d[BrO_3^-]}{dt} = k[Br^-][BrO_3^-][H^+]^2$$

The reaction order with respect to the hydrogen ion is

a. 1 b. 2 c. 6 d. $k[H^+]^2$ e. $k[Br^-][BrO_3^-][H^+]^2$

11. Nitric oxide reacts with hydrogen at a measurable rate at 1000 K according to the equation

$$2NO + 2H_2 \rightarrow N_2 + 2H_2O.$$

The experimental rate law is Rate = $k[NO]^2[H_2]$. When time is in minutes and the concentration is in moles/liter, the units for k are

a. $\frac{\text{moles}}{\ell\text{-min}}$ b. $\frac{\text{-moles}}{\ell\text{-min}}$ c. $\frac{\ell^2}{\text{moles}^2\text{-min}}$ d. $\frac{\text{moles}^2}{\ell^2\text{-min}}$ e. $\frac{\text{moles}^3}{\ell^3\text{-min}}$

12. The exponents in a rate law are determined by

1. the coefficients in the balance equation.
2. experiment.
3. the physical states of the reactants and products.

a. 1 only b. 2 only c. 3 only d. 1 and 2 only e. 1, 2, and 3

13. What is the rate law for the reaction A + B → 2C based on the following kinetic data?

Experiment Number	Initial Conc. (mol/L) $[A]_o$	Initial Conc. (mol/L) $[B]_o$	Initial Rate of Reaction (mol/L·s)
1	0.40	0.10	3.5×10^3
2	0.20	0.10	1.8×10^3
3	0.20	0.50	4.5×10^4

a. rate = $k[A]^1[B]^2$ b. rate = $k[A]^{1/2}[B]^5$ c. rate = $k[A]^2[B]^1$ d. rate = $k[A]^1[B]^{1/5}$

e. rate = $k[A]^{1/2}[B]^2$

14. Kinetic data for the following reaction was determined experimentally:

$$2X + Y \rightarrow 3Z$$

Experiment Number	Initial Conc. (mol/L) $[X]_o$	Initial Conc. (mol/L) $[Y]_o$	Initial Rate of Reaction (mol/L·s)
1	0.15	0.10	6.4×10^2
2	0.15	0.20	2.6×10^3
3	0.30	0.30	1.2×10^4

What is the rate law for the reaction?

a. rate = $k[X]^2[Y]^1$ b. rate = $k[X]^1[Y]^3$ c. rate = $k[2X]^1[Y]^3$ d. rate = $k[X]^1[Y]^2$

e. rate = $k[X]^2[Y]^3$

15. Kinetic data for the following reaction was determined experimentally:

$$A + 2B(g) \rightarrow 2C(g)$$

Experiment Number	Initial Conc. (mol/L) $[A]_o$	Initial Conc. (mol/L) $[B]_o$	Initial Rate of Disappearance of A (mol/L·s)
1	0.15	0.10	4.10×10^3
2	0.45	0.10	3.69×10^4
3	0.15	0.30	1.20×10^4

The rate expression for the above reaction is

a. rate = $k[A]^2[B]$ b. rate = $k[A]^2[2B]^2$ c. rate = $k[A][B]^2$ d. rate = $k[A][2B]$

e. rate = $k[A][2B]^2$

16. For a first order reaction, which of the following can be plotted versus time to give a straight line?

a. ln[A] b. ln kt c. ln[1/A] d. 1/[A] e. [A]

17. For the reaction $2A \rightarrow B$ the reaction is

a. zero-order. b. first-order. c. second-order.

d. impossible to predict without experimental rates at various concentrations of A.

e. impossible to predict without knowing the heat of reaction.

18. For the reaction $2A + 2B \rightarrow 3C$, it was determined that the reaction is third-order overall. The rate law for this reaction might be

a. rate = $k[A]^2[B]^2$ b. rate = $k[A][B]^2$ c. rate = $k[C]^3$ d. rate = $k[A][B]$

e. rate = $k[A][B]^3$

19. A student analyzed a second-order reaction and obtained the graph at the right but forgot to label the axes. What should the labels be for the X and Y coordinates respectively?

a. time, ln[A]

b. time, [A]

c. temperature, [A]

d. temperature, ln[A]

e. time, 1/[A]

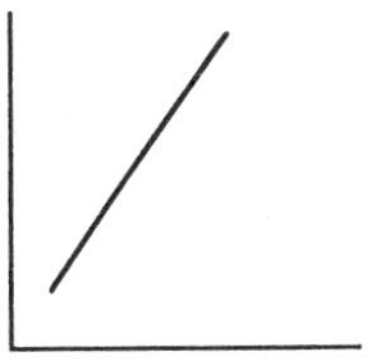

20. What are the units for k, the rate constant, in a first-order reaction where the time unit is seconds (s)?

a. mol/L·s b. mol/L c. 1/s d. s·mol/L e. s·L/mol

21. **Which of the following corresponds to the correct integrated expression for a reaction involving one reactant which is first-order?**

a. $\frac{1}{[A]} = kt + \frac{1}{[A]_o}$ b. k[A] c. $k[A]^2$ d. $\ln[A] = \ln[A]_o - kt$ e. $\frac{1}{[A]_o} - \frac{1}{[A]} = kt$

22. Which of the following corresponds to the correct integrated expression for a reaction involving one reactant which is second-order?

a. $\frac{1}{[A]} = kt + \frac{1}{[A]_o}$ b. k[A] c. $k[A]^2$ d. $\ln[A] = \ln[A]_o - kt$ e. $\frac{1}{[A]_o} - \frac{1}{[A]} = kt$

23. The reaction X → Y follows first-order kinetics with k = 0.83/min. If the initial concentration of X is 3.6 M, what is the concentration of X after 15 minutes?

a. 0.046 M b. 0.230 M c. 1.1×10^{-1} M d. 1.84×10^{-3} M e. 1.4×10^{-5} M

24. The reaction A → B follows first-order kinetics with k = 0.16/min. If the initial concentration of A is 2.4 M, what is the concentration of B after 5.0 minutes?

a. 1.3 M b. 1.1 M c. 0.83 M d. 1.6 M e. 3.4 M

25. What is unique about the half-life of any first-order reaction at 25°C?

a. The units are always sec^{-1}.

b. The value depends only on the rate constant k.

c. The value depends only on the initial concentration of reactant.

d. $\Delta[A_o]/\Delta t = 1$

e. $\Delta[A_o]/\Delta t = 1/2$

26. After five half-life periods for a first-order reaction, what is the molarity of a reagent initially at 0.366 M?

a. 1.14×10^{-2} b. 3.12×10^{-2} c. 6.57×10^{-3} d. 3.12×10^{3} e. 7.32×10^{-2}

27. If the half-life of a first-order process is 3.00 minutes, the rate constant for the process is

a. 1.50/min. b. 1.05/min. c. 4.34/min. d. 0.405/min. e. 0.231/min.

28. Hydrogen peroxide decays into water and oxygen in a first-order process.

$$H_2O_2(aq) \rightarrow H_2O(\ell) + 1/2\ O_2(g)$$

where the rate expression is $-\Delta[H_2O_2]/\Delta t = k[H_2O_2]$. If we begin with 0.100 M H_2O_2 and find that after 3200 seconds, the peroxide concentration falls to 0.0825 M, what is the rate constant, k, at the temperature at which the experiment is performed?

a. 2.61×10^{-5}/s b. 6.01×10^{-5}/s c. 6.59×10^{-5}/s d. 3.79×10^{-4}/s e. 4.24×10^{-3}/s

29. We examine the conversion of cyclopropane to propene, a first-order process. If we begin with 0.0200 M cyclopropane and find that after 168 s the concentration of propene is 0.00286 M, what is the rate constant, k, at this temperature?

a. 2.86×10^{-2}/s b. 4.78×10^{-3}/s c. 3.99×10^{-4}/s d. 9.19×10^{-4}/s e. 8.35×10^{-5}/s

30. Hydrogen peroxide decays into water and oxygen in a first-order process at 20°C

$$H_2O_2(aq) \rightarrow H_2O(\ell) + 1/2\ O_2(g)$$

the rate expression is $-\Delta[H_2O_2]/\Delta t = 1.77 \times 10^{-5}[H_2O_2]$, where k is the units of 1/s. If we begin with 0.0500 M H_2O_2, what will the peroxide concentrations be after exactly two hours?

a. 0.049 M b. 0.044 M c. 0.032 M d. 0.016 M e. 0.0088 M

31. We are studying the reaction of A $\rightarrow$ B and find the following initial rate data:

[A], M	$\Delta[B]/\Delta t$ (M/s)
0.100	0.200
0.200	0.800
0.300	1.80

If we start with a 0.550 M solution of A, what will [A] be after 25 seconds?

a. 0.35 M b. 0.051 M c. 4.0×10^{-3} M d. 2.0×10^{-3} M e. 3.4×10^{-5} M

32. The reaction, $2HI(g) \rightarrow H_2(g) + I_2(g)$, exhibits second-order kinetics $\Delta[HI]/\Delta t = k[HI]^2$ at 400 K. If we begin with $[HI] = 1.00 \times 10^{-3}$ M and find that after 243 sec [HI] drops to 3.56×10^{-4} M, what is the rate constant, k, for the reaction?

a. 1810/M·s b. 7.44/M·s c. 4.48×10^{-3}/M·s d. 5.23×10^{-4}/M·s e. 2.65×10^{-6}/M·s

33. The decomposition of HCO_2H follows first-order kinetics:

$$HCO_2H(g) \rightarrow CO_2(g) + H_2(g)$$

The half-life for the reaction at 550°C is 24 seconds. How many seconds are needed for formic acid, initially 0.82 M, to decrease to 0.018 M?

a. 1.3×10^{2} seconds b. 1.1×10^{3} seconds c. 2.9×10^{3} seconds d. 7.4×10^{3} seconds e. 9.0×10^{4} seconds

34. The half-life for a first-order reaction at 550°C is 85 seconds. How long would it take for 23% of the reactant to decompose?

a. 0.82 seconds b. 26 seconds c. 32 seconds d. 44 seconds e. 180 seconds

35. The decomposition of phosphine, PH_3, follows first-order kinetics:

$$4PH_3(g) \rightarrow P_4(g) + 6H_2(g)$$

The half-life for the reaction at 550°C is 81.3 seconds. How long does it take for the reaction to be 78.5% complete?

a. 8.52 seconds b. 28.4 seconds c. 63.8 seconds d. 117 seconds e. 180 seconds

36. What is the half-life of a first-order reaction if it takes 143 seconds for the concentration to decrease from 1.50 M to 0.0415 M?

a. 0.0251 seconds b. 3.58 seconds c. 4.96 seconds d. 27.6 seconds e. 97.2 seconds

37. What is the half-life of a first-order reaction which is 15% complete after 210 seconds?

a. 7.74 seconds b. 32 seconds c. 76.7 seconds d. 178 seconds e. 895 seconds

38. The Arrhenius equation, $k = Ae^{-Ea/RT}$, may be used to calculate the activation energy from the slope of a line plotted with what parameters?

a. ln k vs. 1/temperature b. ln k vs. 1/time c. 1/k vs. temperature d. 1/k vs. 1/time

e. ln k vs. e^{-T}

39. Which of the following reactions will have the greatest rate at 298 K? Assume that the frequency factor A is the same for all reactions.

a. ΔE = +10 kJ/mol E_a = 25 kJ/mol b. ΔE = -10 kJ/mol E_a = 25 kJ/mol

c. ΔE = -10 kJ/mol E_a = 15 kJ/mol d. ΔE = -10 kJ/mol E_a = 50 kJ/mol

e. ΔE = -10 kJ/mol E_a = 15 kJ/mol

40. In general, as the temperature increases, the rate of a chemical reaction

a. increases due to an increased activation energy.

b. increases only for an endothermic reaction.

c. increases due to a greater number of effective collisions.

d. increases because bonds are weakened.

e. is not changed.

41. If the activation energy for the forward reaction of a given process is +110 kJ and the activation energy for the revenue reaction of the same process is +60.0 kJ, then the energy change for the overall process is

a. -50 kJ b. + 50 kJ c. -170 kJ d. +170 kJ e. -60 kJ

42. In basic solution, $(CH_3)_3CCl$ reacts according to the equation

$$(CH_3)_3CCl + OH^- \rightarrow (CH_3)_3COH + Cl^-$$

The accepted mechanism for the reaction is

$$(CH_3)_3CCl \rightarrow (CH_3)_3C^+ + Cl^- \quad \text{(slow)}$$

$$(CH_3)_3C^+ + OH^- \rightarrow (CH_3)_3COH \quad \text{(fast)}$$

What is the rate law expression for the reaction?

a. rate = k $[(CH_3)_3C^+]^2[OH^-]$ b. rate = k $[(CH_3)_3C^+][OH^-]^2$ c. rate = k $[Cl^-]$

d. rate = k $[(CH_3)_3CCl]$ e. rate = k $[(CH_3)_3CCl][OH^-]$

43. The effect of a catalyst is to

a. increase the number of collisions between reactants.

b. lower the activation energy of a reaction.

c. increase the local temperature at the reactants.

d. decrease the "effective concentrations" of the reactants.

e. increase the energy of the reactants.

44. Calculate the activation energy, E°, for

$$N_2O_5(g) \rightarrow 2NO_2(g) + 1/2\ O_2(g)$$

given k (at 25°C) = 3.46×10^{-5}/s and k (at 50°C) = 1.10×10^{-3}/s.

a. 231 kJ b. 111 kJ c. 99.3 kJ d. 76.2 kJ e. 56.5 kJ

45. The activation energy for $2N_2O(g) \rightarrow 2N_2(g) + O_2(g)$ is 250. kJ. If k for this reaction is 0.380/M·s at 1001 K, what will k be at room temperature, 298 K?

a. 6.36×10^{-32} b. 4.35×10^{-16} c. 0.113 d. 0.216 e. 1.57×10^{31}

46. What is the molecularity of the elementary chemical step

$$H_3C^+ + Cl^- \rightarrow H_3C - Cl?$$

a. unimolecular b. bimolecular c. termolecular d. equimolecular e. first-order

47. Which of the following rate expressions is to be expected for the reaction, $2A + 3B \rightarrow 2E$, given the mechanism below:

Step 1. $A + B \rightarrow C$; slow step

Step 2. $2C \rightarrow D$; fast step

Step 3. $D + B \rightarrow E$; fast step

a. rate = $k[C]^2$ b. rate = $k[A]^2[B]^3$ c. rate = $k[A]^2[B]^2$ d. rate = $k[A][B]$

e. rate = $k[A][B][C]$

48. Which of the following is an appropriate equation for determining the half-life of a first-order reaction?

a. $t_{½} = 0.693/k$ b. $t_{½} = 1/2\ k$ c. $t_{½} = 1/k[A]_o$ d. $t_{½} = [A]_o$ e. $t_{½} = \frac{k}{0.693}$

49. Which of the following can change the rate of a chemical reaction?

1. temperature
2. reactant concentration
3. catalyst

a. 1 only b. 2 only c. 3 only d. 2 and 3 only e. 1, 2, and 3

50. All of the following would be expected to affect the initial rate of a chemical reaction **EXCEPT**

a. adding more reactants.

b. removing some products.

c. decreasing the amount of catalyst.

d. increasing the temperature.

e. decreasing the temperature.

Chapter 15: Answers:

1. b
2. c
3. b
4. b
5. d
6. a
7. c
8. e
9. b
10. b
11. c
12. b
13. a
14. d
15. a
16. a
17. d
18. b
19. e
20. c
21. d
22. a
23. e
24. a
25. b
26. a
27. e
28. b
29. d
30. b
31. d
32. b
33. a
34. c
35. e
36. d
37. e
38. a
39. c
40. c
41. b
42. d
43. b
44. b
45. a
46. b
47. d
48. a
49. e
50. b

Chapter 16
Principles of Reactivity: Chemical Equilibria

1. At equilibrium, what is equal?

   a. concentrations of products and reactants

   b. rate constants for the forward and reserve reactions

   c. the rate of the forward and reverse reaction

   d. the partial pressures of the reactants and products

   e. the reaction quotient and the rate of both reactions

2. In which case does the reaction go farthest to completion?

   a. $K = 10^4$ b. $K = 10^3$ c. $K = 1$ d. $K = 10^{-3}$ e. $K = 10^{-5}$

3. Under which of the following conditions does the equilibrium constant K change for the reaction

$$H_2(g) + I_2(g) \rightleftharpoons 2HI(g)$$

   a. changing the size of the container

   b. introducing more I_2 into the container

   c. measuring the molar concentrations instead of pressures

   d. changing the temperature

   e. none of these, it is always constant

4. Which of the following is true about a chemical system at equilibrium?

   a. no reactions take place

   b. temperature increases will no longer increase reaction rates

   c. the rates of forward and reverse reactions are equal

   d. all reaction products will be solids

   e. it relates concentrations to reactions

5. Which is true about the equilibrium constant expression?

   a. it determines the activation energy needed to perform a reaction

   b. it relates reactant and product concentrations

   c. it relates concentrations to rates

   d. it tells which reactant is highest in concentration

   e. it shows the temperature changes on the equilibrium expression

6. What is the expression for K_c for the following reaction?

$$2NO_2(g) \rightleftharpoons 2NO(g) + O_2(g)$$

a. $K_c = [NO][O_2]/[NO_2]^2$ b. $K_c = [2NO]^2[O_2]/[2NO_2]^2$ c. $K_c = [NO]^2[O_2]/[NO_2]^2$

d. $K_c = [NO_2]^2/[NO]^2[O_2]^2$ e. $K_c = [NO]^2 + [O_2]/[NO_2]^2$

7. What is the equilibrium expression, K_c, for the following reaction?

$$Ca_3(PO_4)_2(s) \rightleftharpoons 3Ca^{2+}(aq) + 2PO_4^{3-}(aq)$$

a. $K_c = [Ca_3(PO_4)_2]/[Ca^{2+}][PO_4^{3-}]^2$ b. $K_c = [Ca^{2+}][PO_4^{3-}]$ c. $K_c = [Ca^{2+}]^3[PO_4^{3-}]^2/[Ca^3(PO_4)_2]$

d. $K_c = [Ca^{2+}]^3[PO_4^{3-}]^2$ e. $K_c = [3Ca^{2+}]^3[2PO_4^{3-}]^2/[Ca_3(PO_4)_2]$

8. For the reaction below, what is the expression for K_c?

$$2H_2(g) + 2FeO(s) \rightleftharpoons 2Fe(s) + 2H_2O(g)$$

a. $K_c = [Fe]^2[H_2O]^2/[H_2]^2[FeO]^2$ b. $K_c = [H_2O]^2/[H_2]^2[FeO]^2$ c. $K_c = [Fe]^2/[H_2]^2[FeO]^2$

d. $K_c = [H_2O]^2/[H_2]^2$ e. $K_c = [2Fe]^2[2H_2O]^2/[2H_2]^2[2FeO]^2$

9. For the reaction below, what is the expression for K_p?

$$SF_6(g) \rightleftharpoons S(s) + 3F_2(g)$$

a. $P_{SF_6}/(P_S \cdot P_{F_2}^3)$ b. $(P_S \cdot P_{F_2}^3)/P$ c. $P_{F_2}^3/P_{SF_6}$ d. $(P_S \cdot 3P_{F_2})^3/P_{SF_6}$ e. $P_{SF_6}/P_{F_2}^3$

10. For the reaction $2A + B_2 \rightleftharpoons C$, $K_c = 1.2$. What is K_c for the reaction $6A + 3B_2 \rightleftharpoons 3C$?

a. 1.1 b. 1.7 c. 3.2 d. 3.6 e. 7.2

11. For the reaction $2A_2 + B \rightleftharpoons 3C$, $K_c = 5.8$. What is K_c for the reaction $3C \rightleftharpoons 2A_2 + B$?

a. 0.17 b. 2.4 c. 2.9 d. 3.6 e. 17.4

12. If $K_c = 0.44$ for the reaction $2NOBr(g) \rightleftharpoons 2NO(g) + Br_2(g)$ at a particular temperature, what is K_c for the following reaction?

$$NOBr(g) \rightleftharpoons NO(g) + 1/2\ Br_2(g)$$

a. 0.19 b. 0.22 c. 0.44 d. 0.66 e. 2.3

13. The equilibrium for the following reaction at 700 K is

$$H_2(g) + I_2(g) \rightleftharpoons 2HI(g) \qquad K_1 = 55.2$$

What is the value of the equilibrium constant for the following reaction (= K_2)?

$$HI(g) \rightleftharpoons 1/2\ H_2(g) + 1/2\ I_2(g) \qquad K_2 = ?$$

a. -55.2 b. 3.29×10^{-5} c. 0.0181 d. 0.135 e. 18.1

14. If $K_c = 1.6 \times 10^{-10}$ at 300°C for the reaction $2SO_3(g) \rightleftharpoons 2SO_2(g) + O_2(g)$ then what is K_c at 300°C for the reaction below?

$$SO_2(g) + 1/2\ O_2(g) \rightleftharpoons SO_3(g)$$

a. 1.3×10^{5} b. 6.25×10^{9} c. 7.9×10^{4} d. 1.6×10^{-10} e. 0.80×10^{-15}

15. If $K_c = 2.5 \times 10^{4}$ at a particular temperature for the reaction

$$2H_2(g) + O_2(g) \rightleftharpoons 2H_2O(g)$$

then what is K_c at the same temperature for the following reaction?

$$H_2O(g) \rightleftharpoons H_2(g) + 1/2\ O_2(g)$$

a. 6.3×10^{-3} b. 1.6×10^{2} c. 1.2×10^{4} d. 1.4×10^{8} e. 6.3×10^{8}

16. Consider the reactions

$$2SO_2(g) + O_2(g) \rightleftharpoons 2SO_3(g) \quad K_1$$

and

$$SO_3(g) \rightleftharpoons SO_2(g) + 1/2\ O_2(g) \quad K_2$$

What is the relationship between the K values of the two reactions?

a. $K_2 = K_1^{1/2}$ b. $K_2 = K_1^{-1/2}$ c. $K_2 = K_1^{2}$ d. $K_2 = -(K_1)^2$ e. $K_2 = 1/2(K_1)^2$

17. Given the following two equilibria,

$$NiCO_3(s) \rightleftharpoons Ni^{2+}(aq) + CO_3^{2-}(aq) \qquad K_1 = 6.6 \times 10^{-9}$$

$$HCO_3^-(aq) + H_2O(\ell) \rightleftharpoons CO_3^{2-}(aq) + H_3O^+(aq) \qquad K_2 = 4.8 \times 10^{-11}$$

calculate the equilibrium constant for the following reaction.

$$NiCO_3(s) + H_3O^+(aq) \rightleftharpoons Ni^{2+}(aq) + HCO_3^-(aq) + H_2O(\ell) \qquad K_3 = ?$$

a. 3.2×10^{-19} b. 1.8×10^{-9} c. 7.3×10^{-3} d. 0.73 e. 140

18. Given the following two equilibria,

$$PbI_2(s) \rightleftharpoons Pb^{2+}(aq) + 2I^-(aq) \qquad K_1 = 8.7 \times 10^{-9}$$

$$PbSO_4(s) \rightleftharpoons Pb^{2+}(aq) + SO_4^{2-}(aq) \qquad K_2 = 1.8 \times 10^{-8}$$

calculate the equilibrium constant for the following reaction.

$$PbSO_4(s) + 2\ I^-(aq) \rightleftharpoons PbI_2(s) + SO_4^{2-}(aq) \qquad K_3 = ?$$

a. 1.6×10^{-16} b. 0.48 c. 2.1 d. 1.2×10^{3} e. 6.4×10^{15}

19. What is the relationship between K_p and K_c for the following reaction?

$$CH_3OH(g) \rightleftharpoons CO(g) + 2H_2(g)$$

a. $K_p = K_c$ b. $K_p = K_c (RT)^{-1}$ c. $K_p = K_c (RT)$ d. $K_p = K_c (RT)^{-2}$ e. $K_p = K_c (RT)^2$

20. What is the relationship between K_p and K_c for the following reaction?

$$2SO_2(g) + O_2(g) \rightleftharpoons 2SO_3(g)$$

a. $K_p = K_c$ b. $K_p = K_c (RT)^{-1}$ c. $K_p = K_c (RT)^{1}$ d. $K_p = K_c (RT)^{2}$ e. $K_p = K_c (RT)^{-2}$

21. In which of the following equations is $K_c = K_p$?

a. $SO_2Cl_2(g) \rightleftharpoons SO_2(g) + Cl_2(g)$

b. $C(s) + H_2O(g) \rightleftharpoons CO(g) + H_2(g)$

c. $2SO_3(g) \rightleftharpoons 2SO_2(g) + O_2(g)$

d. $2HI(g) \rightleftharpoons H_2(g) + I_2(g)$

e. $NO(g) + 1/2Br_2(g) \rightleftharpoons NOBr(g)$

22. We examine the following reaction at 250°C: $PCl_5(g) \rightleftharpoons PCl_3(g) + Cl_2(g)$. At equilibrium we find $[PCl_5] = 3.4 \times 10^{-5}$ M, $[PCl_3] = 1.3 \times 10^{-2}$ M, and $[Cl_2] = 1.0 \times 10^{-4}$ M. Calculate the equilibrium constant, K_c, for the reaction.

a. 26 b. 5.1 c. 0.15 d. 0.038 e. 2.8×10^{-4}

23. A gaseous mixture at 300°C is at equilibrium. It contains 8.7 mol HCl, 0.21 mol H_2 and 0.43 mol Cl_2 in a 1.00 L flask. Calculate K_c for the reaction

$$2HCl(g) \rightleftharpoons H_2(g) + Cl_2(g)$$

a. 96 b. 1.0×10^{-2} c. 1.2×10^{-3} d. 8.4×10^{2} e. 1.1×10^{-4}

24. A gaseous mixture at 500°C is at equilibrium according to the following equation:

$$2NOBr(g) \rightleftharpoons 2NO(g) + Br_2(g)$$

It contains 4.0 mol NOBr, 0.50 mol NO and 0.25 mol Br_2 in a 1.00 L flask. Calculate K_c.

a. 0.31 b. 3.9×10^{-3} c. 9.7×10^{-2} d. 4.1×10^{3} e. 2.4×10^{-4}

25. Consider an equilibrium mixture of oxygen and ozone according to the equation

$$3\,O_2(g) \rightleftharpoons 2\,O_3(g)$$

The partial pressure of O_2 was measured in a flask at equilibrium as 1.25 atm and the total pressure in the flask was 1.75 atm. Calculate K_p. Constant temperature was maintained.

a. 8.0×10^{-3} b. 0.90 c. 0.13 d. 1.6 e. 2.7

26. A chemist prepared a sealed tube with 0.85 atm of PCl_5 at 500 K. The pressure increased as the following reaction occurred. When equilibrium was achieved, the pressure in the tube had increased to 1.25 atm. Calculate K_p.

$$PCl_5(g) \rightleftharpoons PCl_3(g) + Cl_2(g)$$

a. 0.36 b. 0.19 c. 0.10 d. 0.047 e. 0.089

27. A 1.00 liter flask contained 0.24 mol NO_2 at 700 K which decomposed according to the following equation. When equilibrium was achieved, 0.14 mol NO was present. Calculate K_c.

$$2NO_2(g) \rightleftharpoons 2NO(g) + O_2(g)$$

a. 9.6×10^{-3} b. 1.1×10^{-2} c. 9.8×10^{-2} d. 1.4×10^{-1} c. 5.7×10^{3}

28. A mixture of 0.30 mol NO and 0.30 mole CO_2 is placed in a 2.00 L flask and allowed to reach equilibrium at a given temperature. Analysis of the equilibrium mixture indicated that 0.10 mol of CO was present. Calculate K_c for the reaction.

$$NO(g) + CO_2(g) \rightleftharpoons NO_2(g) + CO(g)$$

a. 0.033 b. 0.05 c. 0.25 d. 1.1 e. 0.33

29. We have a gaseous sample where NO_2 and N_2O_4 are in equilibrium. If we know that $[N_2O_4] = 4.6 \times 10^{-5}$, what is $[NO_2]$?

$$2NO_2(g) \rightleftharpoons N_2O_4(g) \qquad K = 180$$

a. 3.8×10^{6} M b. 2.0×10^{3} M c. 5.1×10^{-4} M d. 4.6×10^{-5} M e. 2.6×10^{-7} M

30. We have a gaseious sample where *iso*-butane and *n*-butane are in equilibrium at 300 K. If we know that [*iso*-butane] = 0.040 M, what is [*n*-butane]?

$$n\text{-butane} \rightleftharpoons iso\text{-butane} \qquad K = 2.5$$

a. 0.016 M b. 0.13 M c. 0.73 M d. 2.5 M e. 62.5 M

31. Consider the reaction of carbon disulfide and chlorine according to the following equation:

$$CS_2(g) + 3Cl_2(g) \rightleftharpoons S_2Cl_2(g) + CCl_4(g)$$

When 0.80 mol of CS_2 and 2.4 mol of Cl_2 are placed in a 1.00 liter container and allowed to reach equilibrium at a given temperature, the mixture contains 0.60 mol CCl_4. What is the equilibrium concentration of Cl_2?

a. 0.60 mol/L b. 1.2 mol/L c. 1.6 mol/L d. 1.8 mol/L e. 2.1 mol/L

32. A mixture of 2.0 moles of N_2 and 2.0 moles of O_2 is placed in a 1.00 liter container at a given temperature and allowed to reach equilibrium according to the following equation. Calculate the concentratons of all species present at equilibrium.

$$N_2(g) + O_2(g) \rightleftharpoons 2NO(g) \qquad K_c = 100$$

a. $[N_2] = 0.15$ $[O_2] = 0.15$ $[NO] = 1.5$

b. $[N_2] = 1.67$ $[O_2] = 1.67$ $[NO] = 0.33$

c. $[N_2] = 0.20$ $[O_2] = 0.20$ $[NO] = 2.5$

d. $[N_2] = 0.33$ $[O_2] = 0.33$ $[NO] = 3.3$

e. $[N_2] = 0.33$ $[O_2] = 0.33$ $[NO] = 1.7$

33. A 2.00 liter flask is filled with 1.5 mole SO_3, 2.5 mole SO_2, and 0.5 mole O_2, and allowed to reach equilibrium. At this temperature, $K_c = 1.0$. Predict the effect on the concentration of O_2 as equilibrium is being achieved by using Q, the reaction quotient.

$$2SO_3(g) \rightleftharpoons 2SO_2(g) + O_2(g)$$

a. $[O_2]$ will increase because $Q < K$

b. $[O_2]$ will increase because $Q > K$

c. $[O_2]$ will decrease because $Q < K$

d. $[O_2]$ will decrease because $Q > K$

e. $[O_2]$ will remain the same because $Q = K$

34. Consider the reaction $2A(g) \rightleftharpoons B(g)$ where $K_c = 0.5$ at the temperature of the reaction. If 2.0 moles of A and 2.0 moles of B are introduced into a 1.00 liter flask, what change in concentrations (if any) would occur in time?

a. [A] increases and [B] increases

b. [A] increases and [B] decreases

c. [A] decreases and [B] increases

d. [A] decreases and [B] decreases

e. [A] and [B] remain the same

35. Consider the reaction $A(g) \rightleftharpoons 2B(g)$ where $K_c = 1.5$ at the temperature of the reaction. If 3.0 moles of A and 3.0 moles of B are introduced into a 1.00 liter flask, what change in concentrations (if any) would occur in time?

a. [A] increases and [B] increases

b. [A] increases and [B] decreases

c. [A] decreases and [B] increases

d. [A] decreases and [B] decreases

e. [A] and [B] remain the same

36. At a given temperature, the equilibrium constant $K_c = 1.2$ for the following reaction.

$$2SO_2(g) + O_2(g) \rightleftharpoons 2SO_3(g)$$

If 1.5 mol SO_2, 4.0 mol O_2, and 2.0 mol SO_3 are introduced into a 1.00 liter flask, what changes in concentration (if any) will be observed as the system reaches equilibrium?

a. $[SO_2]$ increases; $[O_2]$ increases; $[SO_3]$ decreases

b. $[SO_2]$ increases; $[O_2]$ decreases; $[SO_3]$ decreases

c. $[SO_2]$ decreases; $[O_2]$ decreases; $[SO_3]$ increases

d. $[SO_2]$ decreases; $[O_2]$ increases; $[SO_3]$ increases

e. all concentrations remain the same

37. Exactly 0.50 mole of sulfur trioxide, 0.10 mole of sulfur dioxide, 0.20 mole of nitrogen monoxide and 0.30 mole nitrogen dioxide are sealed in a 1.0-L flask at 1500°C. The equilibrium constant K_c is 0.24 for the following reaction.

$$SO_3(g) + NO(g) \rightleftharpoons SO_2(g) + NO_2(g) \quad K_c = 0.24$$

When equilibrium is achieved, what changes in concentrations of SO_3 and NO will be observed?

a. $[SO_3]$ increases; [NO] increases

b. $[SO_3]$ increases; [NO] decreases

c. $[SO_3]$ decreases; [NO] decreases

d. $[SO_3]$ decreases; [NO] increases

e. all concentrations remain the same

38. Consider the reaction of iodine and chlorine for which the enthalpy of reaction is -27 kJ.

$$I_2(aq) + Cl_2(aq) \rightleftharpoons 2ICl(g) \qquad \Delta H = -27 \text{ kJ}$$

At 25°C $K_p = 1.6 \times 10^5$. If the temperature is increased to 100°C, what changes (if any) will be observed?

a. K_c will increase b. no change because $K_c = K_p$ c. [ICl] will increase d. $[I_2]$ will increase

e. the partial pressure of ICl will increase

39. In which of the following reactions does a decrease in the volume of the container increase the concentration of the products? Assume constant temperature.

a. $SO_2Cl_2(g) \rightleftharpoons SO_2(g) + Cl_2(g)$

b. $C(s) + H_2O(g) \rightleftharpoons CO(g) + H_2(g)$

c. $2SO_3(g) \rightleftharpoons 2SO_2(g) + O_2(g)$

d. $I_2(g) + Cl_2(g) \rightleftharpoons 2ICl(g)$

e. $2NO(g) + Br_2(g) \rightleftharpoons 2NOBr(g)$

40. The reaction A ⇌ 2B is performed at two temperatures and the equilibrium constant determined. At temperature T_1, the K = 800. At temperature T_2 the K = 20. Which statement is true?

a. $T_1 > T_2$ and the reaction is endothermic.

b. $T_1 < T_2$ and the reaction is endothermic.

c. $T_1 > T_2$ and the reaction is exothermic.

d. T_2 is 40°C less than T_1.

e. T_2 is 400 times greater than T_1.

41. A flask contains the following system at equilibrium:

$$PbCl_2(s) \rightleftharpoons Pb^{2+}(aq) + 2Cl^-(aq)$$

If solid NaCl is added to the system, what change (if any) will be observed?

a. more $PbCl_2$ will be dissolve b. more $PbCl_2$ will precipitate c. more Pb^{2+} will be in solution

d. fewer Cl^- will be in solution e. no change will be observed

42. A flask contains the following system at equilibrium:

$$Mg(OH)_2(s) \rightleftharpoons Mg^{2+}(aq) + 2\ OH^-(aq)$$

Which of the following reagents could be added to increase the solubility of $Mg(OH)_2$?

a. NH_3 b. NaOH c. HCl d. H_2O e. $MgCl_2$

43. Which of the following statements concerning equilibrium is true?

a. Catalysts are an effective means of changing the position of an equilibrium.

b. The concentration of the products equals the concentration of reactions for a reaction at equilibrium.

c. The equilibrium constant may be expressed in pressure terms or concentration terms for any reaction.

d. When two opposing processes are proceeding at the same rate, the system is at equilibrium.

e. A system at equilibrium cannot be disturbed.

44. For the equilibrium system

$$H_2O(g) + CO(g) \rightleftharpoons H_2(g) + CO_2(g) \qquad \Delta H = -42\,\frac{kJ}{mol}$$

K equals 0.62 at 1260 K. If 0.10 mol each of H_2O, CO, H_2, and CO_2 (all at 1260 K) were placed in a 1.0 L thermally insulated vessel which was also at 1260 K, then when the system came to equilibrium

a. the temperature would decrease and the mass of CO would increase.

b. the temperature would decrease and the mass of CO would decrease.

c. the temperature would remain constant and the mass of CO would increase.

d. the temperature would increase and the mass of CO would decrease.

e. the temperature would increase and the mass of CO would increase.

45. For the equilibrium

$$H_2(g) + CO_2(g) \rightleftharpoons H_2O(g) + CO(g) \qquad \Delta H = +42\,\frac{kJ}{mol}$$

K equals 1.6 at 1260 K. If 0.20 mol each of H_2O, CO, H_2, and CO_2 (all at 1260 K) were placed in a 1.0 L thermally insulated vessel which was also at 1260 K, the reaction quotient would have the value of ________ and there would be a net ________ in products and a(n) ________ in the temperature.

	Q	Change in products	Change in temperature
a.	1	increase	increase
b.	1	decrease	decrease
c.	1	increase	decrease
d.	1	decrease	increase
e.	1.6	none	none

46. The process, $2NH_3(g) + \text{heat} \rightleftharpoons 3H_2(g) + N_2(g)$, is at equilibrium at a given temperature. Which of the following changes will change the **equilibrium constant** for the process?

a. Adding more NH_3 to the system (volume held constant).

b. Increasing the pressure by reducing the volume of the system.

c. Adding an effective catalyst for the process.

d. Raising the temperature of the system.

e. Adding some neon to the system.

47. Given the equilibrium

$$PCl_3(g) + Cl_2(g) \rightleftharpoons PCl_5(g) \qquad \Delta H = -92 \text{ kJ}$$

the concentration of Cl_2 at equilibrium will be increased by

a. increasing the pressure.

b. adding PCl_3 to the system.

c. decreasing the temperature.

d. addition of helium.

e. addition of PCl_5.

48. Solid HgO, liquid Hg, and gaseous O_2 are placed in a glass bulb and are allowed to reach equilibrium at a given temperature.

$$2HgO(s) \rightleftharpoons 2Hg(\ell) + O_2(g) \qquad \Delta H = 43.4 \text{ kcal}$$

The mass of HgO in the bulb could be increase by

a. adding more Hg.

b. removing some O_2.

c. reducing the volume of the bulb.

d. increasing the temperature.

e. removing some Hg.

49. We place 0.40 mol of PCl_5 in a 1.0 L flask and let the system come to equilibrium. What will the final concentration of Cl_2 be?

$$PCl_5(g) \rightleftharpoons PCl_3(g) + Cl_2(g) \qquad K_c = 0.47$$

a. 0.40 M b. 0.26 M c. 0.33 M d. 0.18 M e. 0.023 M

50. Ammonium hydrogen sulfide decomposes on heating,

$$NH_4HS(s) \rightleftharpoons NH_3(g) + H_2S(g) \quad K_p = 0.11 \text{ at } 25°C$$

If we have a 1.00 L flask which already contains gaseous NH_3 at a pressure of 0.25 atm and heat up some NH_4HS, what will the equilibrium pressure of NH_3 be at 25°C?

a. .11 atm b. 0.18 atm c. 0.23 atm d. 0.48 atm e. 0.57 atm

Chapter 16: Answers:

1. c
2. a
3. d
4. c
5. b
6. c
7. d
8. d
9. c
10. b
11. a
12. d
13. d
14. c
15. a
16. b
17. e
18. c
19. e
20. b
21. d
22. d
23. c
24. b
25. c
26. a
27. d
28. c
29. c
30. a
31. a
32. d
33. a
34. e
35. b
36. c
37. a
38. d
39. e
40. a
41. b
42. c
43. d
44. a
45. c
46. d
47. e
48. c
49. b
50. d

Chapter 17
Principles of Reactivity: The Chemistry of Acids and Bases

1. Under the Bronsted concept of acids and bases, a base is

a. a proton donor. b. a proton acceptor. c. a hydroxide donor d. an electron pair donor.
e. opposite of a nonelectrolyte.

2. Under the Bronsted concept of acids and bases, an acid is

a. a proton donor. b. a proton acceptor. c. an electron pair donor.
d. an electron pair acceptor. e. a hydroxide ion acceptor.

3. Which of the following is **NOT** an acid-base conjugate pair?

a. HCN and CN^- b. H_2O and OH^- c. H_2S and OH^- d. NH_4^+ and NH_3 e. CH_3COOH and CH_3COO^-

4. In the following reaction

$$HF(aq) + H_2O(\ell) \rightleftharpoons H_3O^+(aq) + F^-(aq)$$

a. HF is an acid and H_3O^+ is its conjugate base.
b. H_2O is an acid and H_3O^+ is its conjugate base.
c. HF is an acid and F^- is its conjugate base.
d. H_2O is an acid and H_3O^+ is its conjugate base.
e. HF is an acid and H_2O is its conjugate base.

5. In the following reaction

$$NH_4^+(aq) + H_2O(\ell) \rightleftharpoons NH_3(aq) + H_3O^+(aq)$$

a. NH_4^+ is an acid and NH_3 is its conjugate base.
b. H_2O is an acid and H_3O^+ is its conjugate base.
c. NH_4^+ is an acid and H_3O^+ is its conjugate base.
d. H_2O is an acid and NH_4^+ is its conjugate base.
e. NH_3 is an acid and NH_4^+ is its conjugate base.

6. In the following reaction

$$HCO_3^-(aq) + H_2O(\ell) \rightleftharpoons CO_3^{2-}(aq) + H_3O^+(aq)$$

a. HCO_3^- is a base and H_3O^+ is its conjugate acid.
b. HCO_3^- is a base and CO_3^{2-} is its conjugate acid.
c. H_2O is a base and H_3O^+ is its conjugate acid.
d. H_2O is a base and CO_3^{2-} is its conjugate acid.
e. H_2O is a base and CO_3^{2-} is its conjugate acid.

7. In the following reaction

$$NO_2^-(aq) + H_2O(\ell) \rightleftharpoons HNO_2(aq) + OH^-(aq)$$

a. H_2O is a base and OH^- is its conjugate acid.

b. H_2O is a base and HNO_2 is its conjugate acid.

c. NO_2^- is a base and HNO_2 is its conjugate acid.

d. NO_2^- is a base and OH^- is its conjugate acid.

e. NO_2^- is a base and H_2O is its conjugate acid.

8. The conjugate acid of OH^- is

a. O^{2-} b. H_2O c. H_3O^+ d. H^+ e. O_2^-

9. Which of the following solutions will have a pH of 1?

a. 1.0 M CH_3CO_2H b. 0.1 M CH_3CO_2H c. 0.1 M HF d. 0.1 M HNO_3 e. 0.1 M NH_3

10. Which of the following is a strong acid?

a. CH_3CO_2H b. HNO_3 c. HNO_2 d. HF e. $B(OH)_3$

11. Which of the following is a weak base?

a. KOH b. $B(OH)_3$ c. NH_3 d. CH_3CO_2H e. $Ba(OH)_2$

12. At 50.0°C the water ionization constant, K_w, is 5.48×10^{-14}. What is the $[H_3O^+]$ in neutral water at 50.0°C?

a. 5.48×10^{-7} b. 2.34×10^{-7} c. 1.00×10^{-7} d. 5.48×10^{-14} e. 4.27×10^{-13}

13. Which of the following solutions will have a pH of 13?

a. 1.0 M NH_3 b. 0.1 M HNO_3 c. 0.1 M CH_3CO_2H d. 0.1 M HF e. 0.1 M KOH

14. All of the following can function both as an acid and base EXCEPT

a. HPO_4^{2-} b. $H_2PO_4^-$ c. HCO_3^- d. OH^- e. CH_3COO^-

15. What is the pH of a 4.2×10^{-4} M HBr solution at 25°C?

a. 2.80 b. 3.38 c. 3.80 d. 4.20 e. 4.62

16. The K_a (HCO_3^-) is the equilibrium constant for the reaction

a. $H_2CO_3 + H_2O \rightleftharpoons H_3O^+ + HCO_3^-$

b. $HCO_3^- + H_2O \rightleftharpoons H_3O^+ + CO_3^{2-}$

c. $HCO_3^- + H_2O \rightleftharpoons H_2CO_3 + OH^-$

d. $HCO_3^- + H_3O^+ \rightleftharpoons H_2CO_3 + H_2O$

e. $HCO_3^- + OH^- \rightleftharpoons CO_3^{2-} + H_2O$

17. The symbol, K_b ($H_2PO_4^-$), is the equilibrium constant for the reaction

a. $H_2PO_4^- + OH^- \rightleftharpoons HPO_4^{2-} + H_2O$

b. $H_2PO_4^- + H_2O \rightleftharpoons HPO_4^{2-} + H_3O^+$

c. $H_2PO_4^- + H_2O \rightleftharpoons H_3PO_4 + OH^-$

d. $H_2PO_4^- + H_3O^+ \rightleftharpoons H_3PO_4 + H_2O$

e. $H_2PO_4^- + 2H_2O \rightleftharpoons 2H_3O_+ + PO_4^{3-}$

18. Which is the strongest acid?

a. Ascorbic acid, $K_a = 8.0 \times 10^{-5}$

b. Benzoic acid, $K_a = 6.5 \times 10^{-5}$

c. 3-chlorobenzoic acid, $K_a = 1.5 \times 10^{-4}$

d. 2-hydroxybenzoic acid, $K_a = 1.1 \times 10^{-3}$

e. Chloroacetic acid, $K_a = 1.4 \times 10^{-3}$

19. Which of the following acids has the strongest conjugate base?

a. Ascorbic acid, $K_a = 8.0 \times 10^{-5}$

b. Benzoic acid, $K_a = 6.5 \times 10^{-5}$

c. 3-chlorobenzoic acid, $K_a = 1.5 \times 10^{-4}$

d. 2-hydroxybenzoic acid, $K_a = 1.1 \times 10^{-3}$

e. Chloroacetic acid, $K_a = 1.4 \times 10^{-3}$

20. Knowing that HF is a stronger acid than H_3CCOOH, determine, if possible, in which direction the following equilibrium lies.

$$HF(aq) + H_3CCOO^-(aq) \rightleftharpoons F^-(aq) + H_3CCOOH(aq)$$

a. equilibrium lies to the left

b. equilibrium lies to the right

c. equilibrium is perfectly balanced left and right

d. can only be determined by only knowing K_a for HF and K_a for CH_3COOH

e. cannot be determined

21. Knowing that H_2S is a stronger acid than HCN, determine, if possible, in which direction the following equilibrium lies.

$$HCN(aq) + HS^-(aq) \rightleftharpoons CN^-(aq) + H_2S(aq)$$

a. equilibrium lies to the left

b. equilibrium lies to the right

c. equilibrium is perfectly balanced left and right

d. can be determined if the relative acidity of HS^- is given

e. cannot be determined

22. What is the pH of a 0.054 M NaOH solution at 25°C?

a. 1.14 b. 1.27 c. 8.64 d. 12.73 e. 13.95

23. What is the pH of a 0.014 M $Ba(OH)_2$ solution at 25°C?

a. 1.54 b. 1.84 c. 10.84 d. 12.16 e. 12.45

24. We have a 0.00100 M solution of NaOH at 25°C. What is $[H_3O^+]$ in this solution?

a. 1.00×10^{-11} M b. 1.00×10^{-9} M c. 1.00×10^{-7} M d. 1.00×10^{-3} M e. 7.00 M

25. We dilute 1.00 mL of 1.00 M HCl solution to 100.0 mL. What is $[OH^-]$ in this solution at 25°C?

a. 1.00×10^{12} M b. 1×10^{2} M c. 0.010 M d. 7.00×10^{-4} M e. 1.00×10^{-12} M

26. We have 500. mL of a solution that contains 0.0854 g of NaOH (MM = 40.0 g/mol). What is the pH of this solution at 25°C?

a. 2.37 b. 2.67 c. 8.54 d. 11.33 e. 11.63

27. We have 300. mL of a solution that contains 0.0128 g of KOH (MM = 56.1 g/mol). What is the pH of this solution at 25°C?

a. 3.12 b. 3.64 c. 10.36 d. 10.88 e. 11.12

28. We have a 4.63×10^{-4} M solution of HCl. What is the pH of this solution at 25°C?

a. 3.33 b. 4.00 c. 4.63 d. 8.37 e. 9.25

29. We have a 0.45 M solution of HNO_3. What is the pH of this solution at 25°C?

a. -0.35 b. 0.35 c. 0.45 d. 3.47 e. 10.53

30. What is $[H_3O^+]$ in a 0.10 M solution of HCN at 25°C? (K_a for HCN = 4.0×10^{-10})

a. 2.0×10^{-5} b. 6.3×10^{-6} c. 4.5×10^{-7} d. 1.6×10^{-9} e. 4.0×10^{-11}

31. What is the pH of a 0.52 M solution of $NaCH_3COO$ at 25°C? (K_b for $CH_3COO^- = 5.6 \times 10^{-10}$)

a. 4.63 b. 4.77 c. 9.23 d. 9.37 e. 10.21

32. What is the pH of a 0.144 M solution of NaF at 25°C? (K_b for $F^- = 1.4 \times 10^{-11}$)

a. 5.85 b. 7.00 c. 7.15 d. 8.15 e. 9.12

33. What is the pH of a 0.49 M solution of NaCN at 25°C? (K_b for $CN^- = 2.5 \times 10^{-5}$)

a. 2.46 b. 5.83 c. 8.17 d. 11.54 e. 12.82

34. A 0.20 M solution of an acid, HA, has a pH of 3.82 at 25°C. What is K_a for this acid?

a. 7.6×10^{-4} b. 4.5×10^{-5} c. 1.1×10^{-7} d. 2.3×10^{-8} e. 4.5×10^{-9}

35. A 0.040 M solution of an acid, HA, has a pH of 3.02 at 25°C. What is K_a for this acid?

a. 2.3×10^{-5} b. 2.6×10^{-5} c. 5.7×10^{-4} d. 2.4×10^{-3} e. 2.4×10^{-2}

36. What is the pH of a 3.18 M CH_3COOH solution at 25°C? $K_a = 1.8 \times 10^{-5}$

a. 2.12 b. 2.75 c. 1.40 d. 4.24 e. 4.74

37. What is the pH of a 1.86 M $CH_3CH_2CO_2H$ solution at 25°C? $K_a = 1.3 \times 10^{-5}$

a. 4.92 b. 4.88 c. 2.42 d. 2.31 e. 2.08

38. What is the pH of a 0.0443 M ammonia (NH_3) solution at 25°C? $K_b = 1.8 \times 10^{-5}$

a. 3.05 b. 6.10 c. 9.25 d. 10.95 e. 12.64

39. What is the pH of a 2.54 M $NH_2(CH_3)$ solution at 25°C? $K_b = 5.0 \times 10^{-4}$

a. 1.45 b. 3.35 c. 10.70 d. 11.10 e. 12.55

40. What is the % ionization of a 3.14 M CH_3CO_2H solution at 25°C? For CH_3CO_2H, $K_a = 1.8 \times 10^{-5}$.

a. 0.24% b. 0.57% c. 1.8% d. 3.2% e. 7.5%

41. At 25°C, what is the pH of a 1.75 M solution of sodium cyanide NaCN? ($K_b = 2.5 \times 10^{-5}$)

a. 11.82 b. 10.04 c. 3.44 d. 2.18 e. 0.80

42. At 25°C, what is the pH of a 3.25 M solution of ammonium chloride, NH_4Cl?

a. 2.37 b. 4.37 c. 4.62 d. 9.37 e. 9.63

43. At 25°C, what is the pH of a 0.084 M solution of potassium nitrite, KNO_2?

a. 4.13 b. 5.87 c. 6.25 d. 8.13 e. 9.87

44. Oxalic acid, $H_2C_2O_4$, is a weak diprotic acid. In a 0.1 M solution of oxalic acid, which species would have the lowest concentration?

a. $H_2C_2O_4$ b. $HC_2O_4^-$ c. H_3O^+ d. $C_2O_4^{2-}$ e. H_2O

45. Under the Lewis concept of acids and bases, a base is

a. a proton donor. b. a proton acceptor. c. an electron pair acceptor.

d. an electron pair donor. e. a hydroxide ion donor

46. Hydrogen sulfide is a weak acid. What is the pH of a 0.10 M H_2S solution? ($K_1 = 1.0 \times 10^{-7}$ and $K_2 = 1.3 \times 10^{-13}$)

a. 1.0 b. 3.5 c. 4.0 d. 5.5 e. 10.4

47. Of the following salts, which one forms a 0.1 M solution with the lowest pH?

a. NH_4Cl b. KCH_3CO_2 c. KBr d. $NaNO_2$ e. NaCl

48. Which of the following salts, when dissolved in water solution, will give the most acidic solution?

a. $MgBr_2$ b. KNO_3 c. LiCl d. Na_2SO_4 e. $FeBr_3$

49. Which of the following compounds will form an acidic aqueous solution?

a. CO_2 b. CaO c. $Mg(OH)_2$ d. NaBr e. $NaCH_3CO_2$

50. All of the following compounds are acids containing chlorine. Which compound is the strongest acid?

a. $HClO_4$ b. $HClO_3$ c. $HClO_2$ d. HClO e. HCl

Chapter 17: answers:

1. b
2. a
3. c
4. c
5. a
6. c
7. c
8. b
9. d
10. b
11. c
12. b
13. e
14. e
15. b
16. b
17. c
18. e
19. b
20. b
21. a
22. d
23. e
24. a
25. e
26. e
27. d
28. a
29. b
30. b
31. c
32. d
33. d
34. c
35. a
36. a
37. d
38. d
39. e
40. a
41. a
42. b
43. d
44. d
45. d
46. c
47. a
48. e
49. a
50. a

Chapter 18
Principles of Reactivity: Reactions Between Acids and Bases

1. What are the products of the following acid-base reaction?

$$NaOH(aq) + HF \rightarrow$$

a. $NaF(aq)$ and $H_2O(\ell)$ b. $NaH(aq)$ and $HOF(aq)$ c. H_2O only d. $H_2O(\ell)$, $F_2(aq)$, and $Na(s)$

e. $NaF(aq) + OH^-(aq)$

2. What are the products of the following acid-base reaction?

$$NaOH(aq) + HNO_3(aq) \rightarrow$$

a. $NaH(aq)$ and $H_2O(\ell)$ b. $NaNO_3(aq)$ and $H_2O(\ell)$ c. H_2O only d. $H_2O(\ell)$, $N_2(g)$, and $Na(s)$

e. $NaNO_3(aq) + H_3O^+$

3. What are the products of the following acid-base reaction?

$$CH_3CO_2H + KOH(aq) \rightarrow$$

a. $CH_3CO_2H(aq)$ and $H_2O(\ell)$ b. $KH(aq)$ and $CH_3CO_2H(aq)$ c. H_2O only

d. $H_2O(\ell)$, $CO_2(g)$, and $K(s)$ e. $KCH_3CO_2(aq)$ and $H_2O(\ell)$

4. Which of the following acid-base reactions will lie predominantly toward the products?

Reaction 1: $HF(aq) + NH_3(aq) \rightleftharpoons NH_4^+(aq) + F^-(aq)$

Reaction 2: $NH_3(aq) + H_2O(\ell) \rightleftharpoons NH_4^+(aq) + OH^-(aq)$

Reaction 3: $HF(aq) + H_2O(\ell) \rightleftharpoons H_3O^+(aq) + F^-(aq)$

a. 1 only b. 2 only c. 1 and 2 only d. 2 and 3 only e. 1, 2, and 3

5. Which of the following acid-base reactions will lie predominantly toward the products?

Reaction 1: $NH_3(aq) + H_2O(\ell) \rightleftharpoons NH_4^+(aq) + OH^-(aq)$

Reaction 2: $CH_3CO_2H(aq) + H_2O(\ell) \rightleftharpoons H_3O^+(aq) + CH_3CO_2^-(aq)$

Reaction 3: $CH_3CO_2H(aq) + NH_3(aq) \rightleftharpoons NH_4^+(aq) + CH_3CO_2^-(aq)$

a. 1 only b. 2 only c. 3 only d. 1 and 2 only e. 1 and 3 only

6. If you mix equal molar quantities of the following substances, how many will produce an acidic solution?

Set 1: $NaOH + HCl$ Set 2: $NaOH + HNO_3$

Set 3: $NH_3 + HCl$ Set 4: $NaOH + CH_3CO_2H$

a. four b. three c. two d. one e. zero (none are acidic)

7. If you mix equal molar quantities of the following substances, how many will produce a neutral solution?

Set 1: $NaOH + HCl$ Set 2: $NaOH + HNO_3$

Set 3: $NH_3 + HCl$ Set 4: $NaOH + CH_3CO_2H$

a. four   b. three   c. two   d. one   e. zero (none are acidic)

8. If you mix equal molar quantities of the following substances, how many will produce a basic solution?

Set 1: $NaOH + HBr$ Set 2: $NaOH + HNO_3$

Set 3: $NH_3 + HF$ Set 4: $NaOH + CH_3CO_2H$

a. four   b. three   c. two   d. one   e. zero (none are acidic)

9. We mix 100. mL of 0.10 M HCl and 100. mL of 0.10 M NaCN. What is the pH of the resulting solution? $K_a(HCN) = 4.0 \times 10^{-10}$

a. 5.20   b. 5.35   c. 6.20   d. 8.65   e. 8.80

10. We mix 100. mL of 0.20 M HBr and 50.0 mL of 0.40 M NaClO. What is the pH of the resulting solution? $K_a(HClO) = 3.5 \times 10^{-8}$

a. 3.92   b. 4.07   c. 4.17   d. 4.86   e. 9.83

11. We have a solution of NH_3. What effect will the addition of HCl have on this solution?

1. increase the pH
2. increase the H_3O^+
3. decrease the pH

a. 1 only   b. 2 only   c. 3 only   d. 1 and 2 only   e. 2 and 3 only

12. We have a solution of NH_4Cl. What effect will addition of NH_3 have on this solution?

1. increase the pH
2. decrease the pH
3. increase the H_3O^+

a. 1 only   b. 2 only   c. 3 only   d. 1 and 3 only   e. 2 and 3 only

13. We have a solution of acetic acid. What effect will the addition of sodium acetate have on this solution?

1. increase the pH
2. decrease the H_3O^+
3. increase the OH^-

a. 1 only   b. 2 only   c. 3 only   d. 1 and 3 only   e. 2 and 3 only

14. We have a solution of NH_4Cl. What effect will the addition of NaCl have on the solution?

1. increase the pH
2. increase the H_3O^+
3. no effect
4. decrease the pH

a. 1 only b. 2 only c. 3 only d. 1 and 2 only e. 2 and 4 only

15. We add 1.00 mL of 10.0 M HNO_3 to 100. mL of 0.10 M NaHCOO. What is the pH of the resulting solution? $K_a(HCOOH) = 1.8 \times 10^{-4}$

a. 2.37 b. 3.45 c. 4.27 d. 4.35 e. 11.60

16. We add 100. mL of 0.10 M NaOH to 100. mL of 0.10 M HClO. What is the pH of the resulting solution? $K_b(ClO^-) = 2.9 \times 10^{-7}$

a. 3.76 b. 3.92 c. 10.08 d. 10.30 e. 10.85

17. We add 200. mL of 0.10 M KOH to 50.0 mL of 0.40 M HF. What is the pH of the resulting solution? $K_b(F^-) = 1.4 \times 10^{-11}$

a. 5.97 b. 6.32 c. 8.02 d. 8.37 e. 9.32

18. If you mix 100. mL of 0.11 M HCl with 50.0 mL of 0.22 M NH_3, what is the pH of the resulting solution? For NH_4^+, $K_a = 5.6 \times 10^{-10}$

a. 4.63 b. 5.19 c. 6.02 d. 8.37 e. 9.37

19. If you mix 250. mL of 0.24 M HF with 75.0 mL of 0.80 M NaOH, what is the pH of the resulting solution? For F^-, $K_b = 1.4 \times 10^{-11}$

a. 5.42 b. 5.79 c. 6.24 d. 7.53 e. 8.21

20. If you mix 50.0 mL of 0.34 M HCN with 100. mL of 0.17 M KOH, what is the pH of the resulting solution?

a. 11.23 b. 10.81 c. 4.60 d. 3.18 e. 2.77

21. If you mix 125. mL of 0.50 M CH_3CO_2H with 75.0 mL of 0.83 M NaOH, what is the pH of the resulting solution? For CH_3COO^-, $K_b = 5.6 \times 10^{-10}$

a. 4.88 b. 5.01 c. 8.99 d. 9.12 e. 9.76

22. If you mix equal molar quantities of NaOH and CH_3CO_2H, what are the principal species present in the resulting solution?

a. Na^+, $CH_3CO_2^-$, OH^-, and H_2O

b. Na^+, $CH_3CO_2^-$, CH_3CO_2H, OH^-, and H_2O

c. Na^+, CH_3CO_2H, OH^-, and H_2O

d. Na^+, $CH_3CO_2^-$, H_3O^+, and H_2O

e. Na^+, CH_3CO_2H, H_3O^+, and H_2O

23. If you mix equal molar quantities of CH_3CO_2H ($K_a = 1.8 \times 10^{-5}$) and NaOH, the resulting solution will be

a. acidic because a small amount of CH_3CO_2H is present.

b. acidic because some excess of H_3O^+ is present.

c. basic because some excess OH^- is present.

d. basic because a small amount of CH_3CO_2H is present.

e. neutral.

24. If you mix equal molar quantities of KOH and HNO_3, the resulting solution will be

a. acidic because a small amount of KNO_3 is present.

b. acidic because some excess H_3O^+ is present.

c. basic because some excess OH^- is present.

d. basic because a small amount of KNO_3 is present.

e. neutral.

25. If you mix equal molar quantities of NH_3 ($K_b = 1.8 \times 10^{-5}$) and CH_3CO_2H ($K_a = 1.8 \times 10^{-5}$), the resulting solution will be

a. acidic because K_a of NH_4^+ is greater than K_b of $CH_3CO_2^-$.

b. acidic because K_a of NH_4^+ is greater than K_a of CH_3CO_2H.

c. basic because K_b of NH_3 is greater than K_b of $CH_3CO_2^-$.

d. basic because K_a of NH_4^+ is greater than K_b of $CH_3CO_2^-$.

e. neutral because K_a of NH_4^+ equals K_b of $CH_3CO_2^-$.

26. If you mix equal molar quantities of NH_3 ($K_b = 1.8 \times 10^{-5}$) and HF ($K_a = 7.2 \times 10^{-4}$), the resulting solution will be

a. acidic because K_a of NH_4^+ is greater than K_b for F^-.

b. acidic because K_a of NH_4^+ is greater than K_a for HF.

c. basic because K_b of NH_4^+ is greater than K_b for F^-.

d. basic because K_a of NH_4^+ is greater than K_b for F^-.

e. neutral.

27. What effect will the addition of the reagent in each of the following have on the pH of the $Ca(OH)_2$ solution respectively?

Flask 1: Addition of HCl to $Ca(OH)_2(aq)$

Flask 2: Addition of NaCl to $Ca(OH)_2(aq)$

a. increase, increase b. decrease, decrease c. decrease, no change d. decrease, increase

e. no change, decrease

28. What effect will the addition of the reagent in each of the following have on the pH of the CH_3CO_2H solution respectively?

Flask 1: Addition of $NaCH_3CO_2$ to $CH_3CO_2H(aq)$

Flask 2: Addition of $Ca(CH_3CO_2)_2$ to $CH_3CO_2H(aq)$

a. no change, increase b. no change, decrease c. decrease, no change d. decrease, decrease

e. increase, increase

29. If you add 0.82 g $NaCH_3CO_2$ (MM = 82.0 g/mol) to 100. mL of a 0.10 M CH_3CO_2H solution, what is the pH of the resulting solution? For CH_3CO_2H, $K_a = 1.8 \times 10^{-5}$

a. 1.73 b. 2.40 c. 3.20 d. 4.74 e. 5.74

30. If you add 0.35 g NaF (MM = 42.0 g/mol) to 150. mL of a 0.30 M HF solution, what is the pH of the resulting solution? For HF, $K_a = 7.2 \times 10^{-4}$

a. 2.41 b. 3.15 c. 3.90 d. 4.21 e. 5.55

31. If you add 0.42 g NH_4Cl (MM = 53.5 g/mol) to 200. mL of 0.34 M NH_3 solution, what is the pH of the resulting solution? For NH_3, $K_b = 1.8 \times 10^{-5}$

a. 3.81 b. 4.74 c. 7.92 d. 9.25 e. 10.19

32. If you add 0.50 g KNO_2 (MM = 85.1 g/mol) to 175 mL of a 0.24 M HNO_2 solution, what is the pH of the resulting solution? For HNO_2, $K_a = 4.5 \times 10^{-4}$

a. 1.81 b. 2.51 c. 3.35 d. 4.72 e. 11.51

33. If you add 1.0 mL of 10.0 M HCl to 500. mL of a 0.10 M NH_3 solution, what is the pH of the resulting solution? For NH_3, $K_b = 1.8 \times 10^{-5}$

a. 4.14 b. 5.80 c. 8.21 d. 9.25 e. 9.86

34. If you add 1.0 mL of 5.0 M NaOH to 250. mL of a 0.35 M CH_3CO_2H solution, what is the pH of the resulting solution? For CH_3COOH, $K_a = 1.8 \times 10^{-5}$

a. 1.54 b. 2.89 c. 3.53 d. 4.74 e. 10.47

35. If you add 20.0 mL of 2.00 M HCl to 250. mL of a 0.42 M $NaCH_3CO_2$ solution, what is the pH of the resulting solution? For CH_3COOH, $K_a = 1.8 \times 10^{-5}$

a. 4.96 b. 5.16 c. 5.32 d. 8.68 e. 9.04

36. If you add 20.0 mL of 2.30 M NH_3 to 100. mL of a 1.17 M NH_4Cl solution, what is the pH of the resulting solution? For NH_3, $K_b = 1.8 \times 10^{-5}$

a. 5.15 b. 6.35 c. 7.10 d. 7.65 e. 8.85

37. If you add 10.0 mL of 2.0 M HCl to 100. mL of a 0.44 M $NaNO_2$ solution, what is the pH of the resulting solution? For HNO_2, $K_a = 4.5 \times 10^{-4}$

a. 3.27 b. 3.43 c. 4.34 d. 10.57 e. 10.73

38. We have 555 mL of a 0.37 M NH_4Cl solution. We add to this solution 1.00 g of pure NaOH(s) (MM = 40.0 g/mol). What is the pH of the resulting solution? $K_a(NH_4^+) = 5.6 \times 10^{-10}$

a. 6.21 b. 8.39 c. 8.94 d. 9.42 e. 10.14

39. We have 300. mL of a 0.20 M solution of HF. How many grams of solid KF (MM = 58.1 g/mol) should be added to make a buffer of pH = 3.00? $K_a(HF) = 7.2 \times 10^{-4}$

a. 0.45 g b. 2.5 g c. 8.4 g d. 9.3 g e. 11 g

40. We have 250. mL of a 0.56 M solution of $NaCH_3COO$. How many milliters of a 0.50 M CH_3COOH solution should be added to make a buffer of pH = 4.40? $K_a(CH_3COOH) = 1.8 \times 10^{-5}$

a. 200 b. 230 c. 620 d. 710 e. 750

41. At the neutralization point of the titration of an acid with base, what condition is met?

a. Volume of base added from buret equals volume acid in reaction flask.

b. Molarity of base from the buret equals molarity of acid in reaction flask.

c. Moles of base added from the buret equals moles of acid in the reaction flask.

d. % ionization of base added from the buret equals % ionization of the acid in flask.

e. All of the above conditions are met.

42. The salt produced by the reaction of an equal number of moles of KOH and HNO_3 will react with water to give a solution which is

a. acidic. b. basic. c. neutral. d. non-ionic. e. impossible to determine.

43. If CH_3CO_2H reacts with an equal number of moles of NaOH, then the pH of the resulting solution is approximately

a. 1. b. 3. c. 7. d. 9. e. 13.

44. What is the pH at the equivalence point in the titration of 10.0 mL of 0.32 M CH_3CO_2H with 15.0 mL of 0.213 M NaOH? For CH_3COO^-, $K_b = 5.6 \times 10^{-10}$

$$CH_3COO^-(aq) + H_2O \rightleftharpoons CH_3COOH(aq) + OH^-(aq)$$

a. 2.62 b. 5.07 c. 7.00 d. 8.93 e. 9.05

45. What is the pH at the equivalence point in the titration of 10.0 mL of 0.16 M NH_3 with 25.0 mL of 0.064 M HCl? For NH_4^+, $K_a = 5.6 \times 10^{-10}$

a. 4.75 b. 5.30 c. 7.00 d. 8.70 e. 9.25

46. What is the pH at the equivalence point in the titration of 20.0 mL of 0.64 M HF with 40.0 mL of 0.32 M NaOH? For F^-, $K_b = 1.4 \times 10^{-11}$

a. 5.07 b. 5.76 c. 8.24 d. 8.93 e. 10.75

47. Which of the following combinations will **NOT** produce a buffer solution?

a. NaCl and CH_3CO_2H b. NH_4Cl and NH_3 c. KCN and HCN d. $NaHCO_3$ and H_2CO_3

e. $NaCH_3CO_2$ and CH_3CO_2H

48. Which of the following combinations can produce a buffer solution?

a. NaCl and CH_3CO_2H b. NH_4Cl and HNO_3 c. KCl and HCN d. $NaCH_3CO_2$ and CH_3CO_2H

e. $NaNO_3$ and HF

49. A buffer solution of pH = 5.30 can be prepared by dissolving acetic acid and sodium acetate in water. How many moles of sodium acetate must be added to 1.0 L of 0.25 M acetic acid to prepare the buffer?

a. 0.28 mol b. 0.47 mol c. 0.90 mol d. 1.8 mol e. 3.6 mol

50. A buffer solution of pH = 9.35 can be prepared by dissolving ammonia and ammonium chloride in water. How many moles of ammonium chloride must be added to 1.0 L of 0.50 M ammonia to prepare the buffer?

a. 0.40 mol b. 0.80 mol c. 1.2 mol d. 2.5 mol e. 4.5 mol

Chapter 18: Answers:

1. a	26. a
2. b	27. c
3. e	28. e
4. a	29. d
5. c	30. a
6. d	31. e
7. c	32. b
8. d	33. e
9. b	34. c
10. c	35. a
11. e	36. e
12. a	37. b
13. d	38. b
14. c	39. b
15. a	40. c
16. c	41. c
17. c	42. c
18. b	43. d
19. e	44. d
20. a	45. b
21. d	46. c
22. b	47. a
23. c	48. d
24. e	49. c
25. e	50. a

Chapter 19
Principles of Reactivity: Precipitation Reactions

1. Which of the following is the solubility product constant for PbI_2?

   a. $K_{sp} = [Pb^{2+}][2I^-]^2$ b. $K_{sp} = [Pb^{2+}][2I^-]$ c. $K_{sp} = [Pb^{2+}]^2[I^-]$ d. $K_{sp} = [Pb^{2+}][I_2^-]^2$

   e. $K_{sp} = [Pb^{2+}][I^-]^2$

2. Which of the following is the solubility product constant for $Ca_3(PO_4)_2$?

   a. $K_{sp} = [Ca^{2+}][PO_4^{3-}]^2$ b. $K_{sp} = [Ca^{2+}]^2[PO_4^{3-}]^3$ c. $K_{sp} = [Ca^{2+}]^3[PO_4^{3-}]^2$ d. $K_{sp} = [Ca^{2+}][PO_4^{3-}]^2$

   e. $K_{sp} = [Ca^{2+}]^6[PO_4^{3-}]^6$

3. Which of the following is the solubility product constant for iron(III) sulfide, Fe_2S_3?

   a. $K_{sp} = [Fe^{2+}][S^{3-}]$ b. $K_{sp} = [Fe^{3+}][S^{2-}]^3$ c. $K_{sp} = [Fe^{2+}]^2[S^{3-}]^3$ d. $K_{sp} = [Fe^{3+}]^2[S^{2-}]^3$

   e. $K_{sp} = [Fe^{2+}][S^{2-}]^3$

4. Which of the following is the solubility product constant for $Mn(OH)_2$?

   a. $K_{sp} = [Mn^{2+}][OH^-]^2$ b. $K_{sp} = [Mn^{2+}][2\ OH^-]^2$ c. $K_{sp} = [Mn^{2+}]^2[OH^-]^2$ d. $K_{sp} = [Mn^{2+}]^2[OH^-]$

   e. $K_{sp} = [Mn^{2+}]^2[2\ OH^-]^2$

5. The solubility of $FeCO_3$ is 5.9×10^{-6} mol/L. What is K_{sp} for $FeCO_3$?

   a. 5.9×10^{-6} b. 1.2×10^{-21} c. 3.5×10^{-11} d. 2.8×10^{-10} e. 1.3×10^{-14}

6. The solubility of HgS is 5.5×10^{-27} mol/L. What is K_{sp} for HgS?

   a. 4.0×10^{-3} b. 8.2×10^{-4} c. 1.3×10^{-13} d. 7.4×10^{-14} e. 3.0×10^{-53}

7. The solubility of $PbBr_2$ is 0.0116 M. What is K_{sp} for $PbBr_2$?

   a. 1.6×10^{-6} b. 6.2×10^{-6} c. 1.3×10^{-4} d. 3.1×10^{-6} e. 1.1×10^{-1}

8. The solubility of Ag_2SO_4 is 0.0162 M. What is K_{sp} for Ag_2SO_4?

   a. 1.7×10^{-5} b. 6.1×10^{-3} c. 2.6×10^{-4} d. 1.4×10^{-4} e. 5.2×10^{-4}

9. Which of the following expressions describes the relationship between the solubility product, K_{sp}, and the solubility, s, of MgF_2?

   a. $K_{sp} = 2s$ b. $K_{sp} = s^2$ c. $K_{sp} = 2s^3$ d. $K_{sp} = 4s^2$ e. $K_{sp} = 4s^3$

10. Which of the following compounds has the highest molar solubility?

a. AgCl; $K_{sp} = 1.8 \times 10^{-10}$ b. CuBr; $K_{sp} = 5.3 \times 10^{-9}$ c. AgBr; $K_{sp} = 3.3 \times 10^{-13}$
d. CuI; $K_{sp} = 5.1 \times 10^{-12}$ e. CuCl; $K_{sp} = 1.9 \times 10^{-7}$

11. Calculate the molar solubility of BiI_3. $K_{sp} = 8.1 \times 10^{-19}$

a. 3.0×10^{-5} M b. 1.3×10^{-5} M c. 5.9×10^{-7} M d. 4.5×10^{-8} M e. 6.8×10^{-11} M

12. Calculate the molar solubility of Fe_2S_3. $K_{sp} = 1.4 \times 10^{-88}$

a. 5.5×10^{-62} M b. 1.2×10^{-44} M c. 5.2×10^{-30} M d. 4.8×10^{-24} M e. 1.1×10^{-18} M

13. Which of the following compounds has the highest molar solubility?

a. $BaSO_4$; $K_{sp} = 1.1 \times 10^{-10}$ b. $FeCO_3$; $K_{sp} = 3.5 \times 10^{-11}$ c. $PbSO_4$; $K_{sp} = 1.8 \times 10^{-8}$
d. $SrCO_3$; $K_{sp} = 9.4 \times 10^{-10}$ e. $ZnCO_3$; $K_{sp} = 1.5 \times 10^{-11}$

14. Which of the following compounds has the lowest molar solubility?

a. $BaSO_4$; $K_{sp} = 1.1 \times 10^{-10}$ b. $FeCO_3$; $K_{sp} = 3.5 \times 10^{-11}$ c. $PbSO_4$; $K_{sp} = 1.8 \times 10^{-8}$
d. $SrCO_3$; $K_{sp} = 9.4 \times 10^{-10}$ e. $ZnCO_3$; $K_{sp} = 1.5 \times 10^{-11}$

15. Rank the compounds from lowest to highest molar solubility.

$FeCO_3$; $K_{sp} = 3.5 \times 10^{-11}$
$BaSO_4$; $K_{sp} = 1.1 \times 10^{-10}$
$ZnCO_3$; $K_{sp} = 1.5 \times 10^{-11}$

a. $ZnCO_3 < BaSO_4 < FeCO_3$ b. $FeCO_3 < ZnCO_3 < BaSO_4$ c. $ZnCO_3 < FeCO_3 < BaSO_4$
d. $BaSO_4 < ZnCO_3 < FeCO_3$ e. $BaSO_4 < FeCO_3 < ZnCO_3$

16. Which of the following compounds has the highest molar solubility?

a. $PbCO_3$; $K_{sp} = 1.5 \times 10^{-13}$ b. PbS; $K_{sp} = 8.4 \times 10^{-28}$ c. $PbSO_4$; $K_{sp} = 1.8 \times 10^{-8}$
d. $PbCrO_4$; $K_{sp} = 1.8 \times 10^{-14}$ e. PbI_2; $K_{sp} = 8.7 \times 10^{-9}$

17. Which of the following will give a saturated solution with the highest concentration of iodide ion, I^-?

a. CuI; $K_{sp} = 5.1 \times 10^{-12}$ b. AuI; $K_{sp} = 1.6 \times 10^{-23}$ c. AgI; $K_{sp} = 1.5 \times 10^{-16}$
d. BiI_3; $K_{sp} = 8.1 \times 10^{-19}$ e. AuI_3; $K_{sp} = 1.0 \times 10^{-46}$

18. What is the concentration of SO_4^{2-} in a saturated solution of $BaSO_4$ if $K_{sp} = 1.1 \times 10^{-10}$?

a. 1.1×10^{-10} M b. 5.5×10^{-11} M c. 5.0×10^{-5} M d. 1.0×10^{-5} M e. 9.5×10^{-4} M

19. What is the concentration of CrO_4^{2-} in a saturated solution of $PbCrO_4$ if $K_{sp} = 1.8 \times 10^{-14}$?

a. 1.3×10^{-7} M b. 7.5×10^{-6} M c. 1.8×10^{-4} M d. 1.3×10^{-4} M e. 5.1×10^{-3} M

20. What is the concentration of F^- in a saturated solution of BaF_2 if $K_{sp} = 1.7 \times 10^{-6}$?

a. 7.5×10^{-3} M b. 8.2×10^{-4} M c. 1.5×10^{-2} M d. 4.3×10^{-7} M e. 1.5×10^{-6} M

21. Which of the following has the highest molar solubility?

a. $PbCO_3$; $K_{sp} = 1.5 \times 10^{-13}$ b. PbS; $K_{sp} = 8.4 \times 10^{-28}$ c. PbI_2; $K_{sp} = 8.7 \times 10^{-9}$

d. $PbSO_4$; $K_{sp} = 1.8 \times 10^{-8}$ e. $Pb_3(PO_4)_2$; $K_{sp} = 3.0 \times 10^{-44}$

22. If excess AgCl is placed in 100.0 mL water, what is the chloride ion concentration in the solution? $K_{sp} = 1.8 \times 10^{-10}$

a. 4.8×10^{-3} M b. 9.6×10^{-3} M c. 1.0×10^{-4} M d. 1.3×10^{-5} M e. 1.8×10^{-10} M

23. If excess $BaCO_3$ is placed in 250. mL water, what is the barium ion concentration in the solution? $K_{sp} = 8.1 \times 10^{-9}$

a. 3.0×10^{-3} M b. 6.0×10^{-4} M c. 9.0×10^{-5} M d. 1.5×10^{-6} M e. 7.6×10^{-6} M

24. Calculate the equilibrium constant for the reaction

$$CuCl(s) + I^-(aq) \rightleftharpoons CuI(s) + Cl^-(aq)$$

CuCl; $K_{sp} = 1.9 \times 10^{-7}$ CuI; $K_{sp} = 5.1 \times 10^{-12}$

a. 8.4×10^{-2} b. 2.3×10^{-6} c. 3.7×10^{4} d. 4.4×10^{17} e. 9.7×10^{-19}

25. Calculate the equilibrium constant for the reaction:

$$CdS(s) + Zn^{2+}(aq) \rightleftharpoons ZnS(s) + Cd^{2+}(aq)$$

CdS; $K_{sp} = 3.6 \times 10^{-29}$ ZnS; $K_{sp} = 1.1 \times 10^{-21}$

a. 3.3×10^{-8} b. 2.7×10^{-4} c. 4.2×10^{5} d. 2.5×10^{49} e. 3.1×10^{7}

26. For AgI, $K_{sp} = 1.5 \times 10^{-16}$. If you mix 500. mL of 1×10^{-8} M $AgNO_3$ and 500. mL of 1×10^{-8} M NaI, what will be observed?

a. A precipitate forms because $Q_{sp} > K_{sp}$.

b. A precipitate forms because $Q_{sp} < K_{sp}$.

c. No precipitate forms because $Q_{sp} = K_{sp}$.

d. No precipitate forms because $Q_{sp} < K_{sp}$.

e. No precipitate forms because $Q_{sp} > K_{sp}$.

27. For $BaSO_4$, $K_{sp} = 1.1 \times 10^{-10}$. If you mix 200. mL of 1.0×10^{-4} M $Ba(NO_3)_2$ and 500. mL of 8.0×10^{-2} M H_2SO_4, what will be observed?

a. A precipitate forms because $Q_{sp} > K_{sp}$.

b. A precipitate forms because $Q_{sp} < K_{sp}$.

c. No precipitate forms because $Q_{sp} = K_{sp}$.

d. No precipitate forms because $Q_{sp} < K_{sp}$.

e. No precipitate forms because $Q_{sp} > K_{sp}$.

28. For MgF_2, $K_{sp} = 6.4 \times 10^{-9}$. If you mix 400. mL of 1×10^{-4} M $Mg(NO_3)_2$ and 500. mL of 1.00×10^{-4} M NaF, what will be observed?

a. A precipitate forms because $Q_{sp} > K_{sp}$.

b. A precipitate forms because $Q_{sp} < K_{sp}$.

c. No precipitate forms because $Q_{sp} = K_{sp}$.

d. No precipitate forms because $Q_{sp} < K_{sp}$.

e. No precipitate forms because $Q_{sp} > K_{sp}$.

29. For $ZnCO_3$, $K_{sp} = 1.5 \times 10^{-11}$. If you mix 250. mL of 2.0×10^{-3} M $ZnCl_2$ and 750. mL of 4.00×10^{-8} M $CaCO_3$, what will be observed?

a. A precipitate forms because $Q_{sp} > K_{sp}$.

b. A precipitate forms because $Q_{sp} < K_{sp}$.

c. No precipitate forms because $Q_{sp} = K_{sp}$.

d. No precipitate forms because $Q_{sp} < K_{sp}$.

e. No precipitate forms because $Q_{sp} > K_{sp}$.

30. A saturated solution of $Ca(OH)_2$ has a pH = 12.40. What is K_{sp} for $Ca(OH)_2$?

a. 2.5×10^{-2} b. 1.3×10^{-2} c. 8.0×10^{-6} d. 2.0×10^{-6} e. 4.0×10^{-13}

31. A saturated solution of cobalt hydroxide, $Co(OH)_2$ has a pH = 8.90. What is K_{sp} for $Co(OH)_2$?

a. 2.1×10^{-27} b. 2.5×10^{-16} c. 5.0×10^{-16} d. 4.0×10^{-12} e. 1.3×10^{-2}

32. Calculate the molar solubility of AgCl in a 0.10 M solution of NaCl. (K_{sp} of AgCl is 1.8×10^{-10}.)

a. 1.3×10^{-5} M b. 5.5×10^{8} M c. 1.8×10^{-9} M d. 4.2×10^{-5} M e. 4.8×10^{-4} M

33. Calculate the molar solubility of PbI_2 in a 0.40 M solution of NaI. (K_{sp} of PbI_2 is 8.7×10^{-8}.)

a. 4.4×10^{-3} M b. 2.8×10^{-3} M c. 2.8×10^{-4} M d. 5.4×10^{-7} M e. 2.3×10^{-7} M

34. What is the molar solubility of Fe^{3+} in a solution that is buffered at a pH of 5.00? (K_{sp} of $Fe(OH)_3$ is 6.3×10^{-38}.)

a. 6.3×10^{-23} M b. 4.2×10^{-12} M c. 6.3×10^{-11} M d. 1.2×10^{-8} M e. 4.2×10^{-8} M

35. For $BaSO_4$, $K_{sp} = 1.1 \times 10^{-10}$. What is the molar solubility of $BaSO_4$ in a solution which is 0.018 M in Na_2SO_4?

a. 0.018 mol/L b. 7.8×10^{-5} mol/L c. 1.1×10^{-5} mol/L d. 6.1×10^{-9} mol/L
e. 1.1×10^{-10} mol/L

36. For AgI, $K_{sp} = 8.3 \times 10^{-17}$. What is the molar solubility of AgI in a solution which is 5.1×10^{-4} M in $AgNO_3$?

a. 5.1×10^{-2} mol/L b. 1.1×10^{-5} mol/L c. 8.3×10^{-11} mol/L d. 1.6×10^{-13} mol/L
e. 4.2×10^{-20} mol/L

37. For $BaSO_4$, $K_{sp} = 1.1 \times 10^{-10}$. If 1.00 g of $BaSO_4$ (MM = 233.5 g/mol) is placed in 1.00 L of pure water at 25°C, how much of it will dissolve?

a. All of it. b. 0.35 g c. 0.0024 g d. 4.0×10^{-2} g e. 1.5×10^{-4} g

38. For $NiCO_3$, $K_{sp} = 6.6 \times 10^{-9}$. If 0.050 g of $NiCO_3$ (MM = 118.7 g/mol) is placed in 1.00 L of pure water at 25°C, how much of it will dissolve?

a. All of it. b. 0.044 g c. 0.037 g d. 0.025 g e. 0.0096 g

39. For CaF_2, $K_{sp} = 3.9 \times 10^{-11}$. If 0.0080 g of CaF_2 (MM = 58.1 g/mol) is placed in 1.00 L of pure water at 25°C, how much of it will dissolve?

a. All of it. b. 0.00022 g c. 0.0040 g d. 0.0025 e. 0.0079 g

40. For Ag_2SO_4, $K_{sp} = 1.7 \times 10^{-5}$. If 1.00 g of Ag_2SO_4 (MM = 311.8 g/mol) is placed in 2.00 L of pure water at 25°C, how much of it will dissolve?

a. All of it. b. 0.94 g c. 0.47 g d. 0.18 g e. 0.014 g

41. For Ag_2SO_4, $K_{sp} = 1.7 \times 10^{-5}$. How many grams of Na_2SO_4 (MM = 142.0 g/mol) must be added to 100. mL of 0.022 M $AgNO_3$ to just initiate precipitation?

a. 5.0 g b. 4.0 g c. 3.0 g d. 0.50 g e. 0.40 g

42. For thallium bromide, TlBr, $K_{sp} = 3.4 \times 10^{-6}$. How many grams of KBr (MM = 119.0 g/mol) must be added to 100. mL of 5.5×10^{-4} M $TlNO_3$ to just initiate precipitation?

a. 0.74 g b. 0.074 g c. 0.065 g d. 0.0065 g e. 0.0033 g

43. For $PbCl_2$, $K_{sp} = 1.7 \times 10^{-5}$. How many grams of NaCl (MM = 58.44) must be added to 200. mL of 0.16 M $Pb(NO_3)_2$ to just initiate precipitation?

a. 2.34 g b. 0.85 g c. 0.60 g d. 0.43 g e. 0.12 g

44. For $CaCrO_4$, $K_{sp} = 7.1 \times 10^{-4}$. How many grams of Na_2CrO_4 (MM = 162.0 g/mol) must be added to 200. mL of 0.250 M $Ca(NO_3)_2$ to just initiate precipitation?

a. 0.046 g b. 0.092 g c. 0.34 g d. 1.8 g e. 2.1 g

45. The solubility of salts can be affected by other equilibria. How is the solubility of $ZnCO_3$ changed by the following reaction?

$$CO_3^{2-}(aq) + H_2O(\ell) \rightleftharpoons HCO_3^-(aq) + OH^-(aq)$$

a. The solubility is increased because the CO_3^{2-} undergoes further reaction.

b. The solubility is decreased because the CO_3^{2-} undergoes further reaction.

c. The solubility is increased because a gas is formed.

d. The solubility is decreased because a gas is formed.

e. The solubility stays the same because a saturated $ZnCO_3$ solution is at equilibrium.

46. The solubility of salts can be affected by other equilibria. How is the solubility of PbS changed by the following reaction?

$$S^{2-}(aq) + H_2O(\ell) \rightleftharpoons HS^-(aq) + OH^-(aq)$$

a. The solubility is increased because a gas is formed.

d. The solubility is decreased because a gas is formed.

c. The solubility stays the same because the solution is saturated.

d. The solubility is increased because the S^{2-} undergoes further reaction.

e. The solubility is decreased because the S^{2-} undergoes further reaction.

47. The solubility of salts can be affected by other equilibria. Which of the following will affect the solubility of ZnS?

1. Hydrolysis of $S^{2-}(aq)$
2. Addition of $H^+(aq)$
3. Addition of $OH^-(aq)$

a. 1 only b. 2 only c. 3 only d. 1 and 2 only e. 1, 2, and 3

48. The solubility of salts can be affected by other equilibria. Addition of all the following will affect the solubility of $FeCO_3$ **EXCEPT**

a. $NaHCO_3$ b. NaCl c. H_2CO_3 d. Na_2CO_3 e. HCl

49. We can dissolve silver chloride in aqueous ammonia.

$AgCl(s) + 2NH_3(aq) \rightleftharpoons [Ag(NH_3)_2]^+(aq) + Cl^-(aq)$ $K_{dissolve} = ?$

This process is actually the sum of two others:

$AgCl(s) \rightleftharpoons Ag^+(aq) + Cl^-(aq)$ $K_1 = 1.8 \times 10^{-10}$

$Ag^+(aq) + 2NH_3(aq) \rightleftharpoons [Ag(NH_3)_2]^+(aq)$ $K_2 = 1.6 \times 10^7$

The value of $K_{dissolve}$, the equilibrium constant for the overall process, is

a. 1.6×10^7 b. 3.5×10^2 c. 2.9×10^{-3} d. 4.2×10^{-8} e. 1.1×10^{-17}

50. The cations Fe^{3+}, Pb^{2+}, and Al^{3+} can be precipitated as insoluble hydroxides: $Fe(OH)_3$, $Pb(OH)_2$, and $Al(OH)_3$. If you have a solution containing these three cations, each with a concentration of 0.10 M, what is the order in which they precipitate as hydroxides?

Compound	K_{sp}
$Fe(OH)_3$	6.3×10^{-38}
$Pb(OH)_2$	2.8×10^{-16}
$Al(OH)_3$	1.9×10^{-33}

a. $Fe(OH)_3$ first, then $Al(OH)_3$, finally $Pb(OH)_2$

b. $Al(OH)_3$ first, then $Fe(OH)_3$, finally $Pb(OH)_2$

c. $Pb(OH)_2$ first, then $Al(OH)_3$, finally $Fe(OH)_3$

d. $Fe(OH)_3$ first, then $Pb(OH)_2$, finally $Al(OH)_3$

e. $Al(OH)_3$ first, then $Pb(OH)_2$, finally $Fe(OH)_3$

Chapter 19: Answers:

1. e
2. c
3. d
4. a
5. c
6. e
7. b
8. a
9. e
10. e
11. b
12. e
13. c
14. e
15. c
16. e
17. d
18. d
19. a
20. c
21. c
22. d
23. c
24. c
25. a
26. d
27. a
28. d
29. c
30. c
31. b
32. c
33. d
34. c
35. d
36. d
37. c
38. e
39. a
40. a
41. d
42. b
43. e
44. b
45. a
46. d
47. e
48. b
49. c
50. a

Chapter 20
Principles of Reactivity: Entropy and Free Energy

1. If the reaction A + B $\rightarrow$ C has an equilibrium constant less than one, which of the following statements is true?

   a. The reaction is non-spontaneous. b. The reaction is spontaneous. c. The reaction will not occur.
   d. The reaction will happen instantly. e. The reaction will explode.

2. The disorder of a system is represented by the

   a. enthalpy. b. Gibbs free energy. c. entropy. d. heat of vaporization.
   e. equilibrium constant.

3. Which of the following represents an increase in entropy?

   a. freezing of water b. boiling of water c. crystallization of salt from a supersaturated solution
   d. the reaction $2NO(g) \rightarrow N_2O_2(g)$ e. the reaction $2H_2(g) + O_2(g) \rightarrow 2H_2O(g)$

4. The enthalpy of vaporization of methanol (CH_3OH) is 35.3 kJ/mol at the boiling point of 64.2°C. Calculate the entropy change for methanol going from a liquid to vapor.

   a. 600. J/K·mol b. 551 J/K·mol c. 105 J/K·mol d. -105 J/K·mol e. -551 J/K·mol

5. The change in entropy for the vaporization of CCl_4 is +85.7 J/K·mol and the boiling temperature is 77°C. What is the heat of vaporization?

   a. 1110 kJ/mol b. 245 kJ/mol c. 30.0 kJ/mol d. 4.08 kJ/mol e. -245 kJ/mol

6. Calculate the change in entropy for the condensation of butane, the fuel in hand-held, disposable lighters at its boiling point of -0.5°C given ΔH_{vap} = 24.3 kJ/mol.

   a. -89.2 J/K·mol b. -0.0892 J/K·mol c. 0.0892 J/K·mol d. 89.2 J/K·mol e. 53600 J/K·mol

7. Calculate the standard entropy change for the following reaction,

$$Cu(s) + 1/2\ O_2(g) \rightarrow CuO(s)$$

   given that S°[Cu(s)] = 33.15 J/K·mol, S°[O_2(g)] = 205.14 J/K·mol, and S°[CuO(s)] = 42.63 J/K·mol

   a. 195.66 J/K b. 93.09 J/K c. -45.28 J/K d. -93.09 J/K e. 195.66 J/K

8. Calculate the standard entropy change for the following reaction,

$$CCl_4(\ell) + O_2(g) \rightarrow CO_2(g) + 2Cl_2(g)$$

given that S°[$CCl_4(\ell)$] = 216.40 J/K·mol, S°[$CO_2(g)$] = 213.74 J/K·mol, S°[$O_2(g)$] = 205.14 J/K·mol, and S°[$Cl_2(g)$] = 223.07 J/K·mol.

a. -25.78 J/K b. -15.27 J/K c. +1.93 J/K d. 238.34 J/K e. 317.42 J/K

9. Calculate the standard entropy change for the following reaction,

$$CH_4(g) + 2\,O_2(g) \rightarrow CO_2(g) + 2H_2O(\ell)$$

given that S°[$CO_2(g)$] = 213.74 J/K·mol, S°[$O_2(g)$] = 205.14 J/K·mol, S°[$H_2O(\ell)$] = 69.91 J/K·mol, and S°[$CH_4(g)$] = 186.26 J/K·mol.

a. -312.89 J/K b. -242.98 J/K c. -118.42 J/K d. 23.5 J/K e. 312.89 J/K

10. Calculate the standard molar entropy of urea ($CO(NH_2)_2(s)$ if the standard entropy change for the formation is -456.3 J/K·mol and given S°[C(s)] = 5.74 J/K·mol, S°[$O_2(g)$] = 205.1 J/K·mol, S°[$N_2(g)$] = 191.6 J/K·mol, and S°[$H_2(g)$] = 130.7 J/K·mol.

a. -1017.2 J/K·mol b. +314.1 J/K·mol c. +194.2 J/K·mol d. +105.0 J/K·mol
e. -56.0 J/K·mol

11. The entropy of a substance ______ increases as it changes from a liquid to a gas.

a. never b. sometimes c. always d. often e. rarely

12. Entropy generally __________ with __________ molecular complexity.

a. increases, increasing b. decreases, increasing c. decreases, decreasing
d. increases, decreasing e. dissipates, decreasing

13. Which of the following do you expect to have the largest entropy at 25°C?

a. Fe(s) b. Xe(g) c. $H_2O(\ell)$ d. $Hg(\ell)$ e. He(g)

14. Which of the following do you expect to have the largest entropy at 25°C?

a. $H_2O(\ell)$ b. $H_2O(s)$ c. $H_2O(g)$ d. $O_2(g)$ e. $CCl_4(g)$

15. In which of the following reactions do you expect to have a decrease in entropy?

a. $Fe(s) \rightarrow Fe(\ell)$ b. $Fe(s) + S(s) \rightarrow FeS(s)$ c. $2Fe(s) + 3/2\,O_2(g) \rightarrow Fe_2O_3(s)$
d. $HF(\ell) \rightarrow HF(g)$ e. $2H_2O_2(\ell) \rightarrow 2H_2O(\ell) + O_2(g)$

16. In which of the following reactions do you expect to have the largest increase in entropy?

a. $I_2(s) \rightarrow I_2(g)$ b. $2IF(g) \rightarrow I_2(g) + F_2(g)$ c. $Mn(s) + O_2(g) \rightarrow MnO_2(s)$

d. $Hg(\ell) + S(s) \rightarrow HgS(s)$ e. $CuSO_4(s) + 5H_2O(\ell) \rightarrow CuSO_4{\cdot}5H_2O(s)$

17. In which of the following reactions do you expect to have the smallest entropy change?

a. $2HF(g) + Cl_2(g) \rightarrow 2HCl(g) + F_2(g)$ b. $2Fe(s) + 3/2\ O_2(g) \rightarrow Fe_2O_3(s)$

c. $CH_4(g) + 2\ O_2(g) \rightarrow CO_2(g) + 2H_2O(\ell)$ d. $Cu(s) + 1/2\ O_2(g) \rightarrow CuO(s)$

e. $H_2(g) + I_2(g) \rightarrow 2HI(g)$

18. The formation

$$1/2\ A_2 + 2B_2 + C \rightarrow CAB_4$$

has an enthalpy of formation of -104 kJ and a change in entropy of -60.8 J/K at 30°C. What is ΔG and spontaneity of the reaction?

a. -85.6 kJ, spontaneous b. -18.3 kJ, not spontaneous c. +18.3 kJ, spontaneous

d. +85.6 kJ, not spontaneous e. -85.6 kJ, not spontaneous

19. If ΔG is negative at all temperatures, then ΔS is _________ and ΔH is _________.

a. positive, negative b. zero, large c. negative, positive d. large, zero e. small, large

20. If ΔG is positive at all temperatures, then ΔS is _________ and ΔH is _________.

a. positive, negative b. negative, positive c. small, zero d. large, zero e. large, small

21. If ΔH and ΔS are both negative or positive, then ΔG has a _________ sign.

a. positive b. negative c. variable d. large e. no

22. At what temperature would a given reaction become spontaneous if ΔH = +119 kJ and ΔS = +263 J/K?

a. 452 K b. 2210 K c. 382 K d. 2.21 K e. 363 K

23. The free energy change for a given reaction is -36.2 kJ. What is the equilibrium constant at 298 K?

a. 0.985 b. 2.22×10^6 c. 1.01 d. 8.32×10^{-7} e. 3.25×10^6

24. The formation constant for the formation of the complex ion AlF_6^{3-} is 5.0×10^{23}. What is the standard free energy change for the process?

a. -140 kJ b. -59 kJ c. 59 kJ d. 140 kJ e. -2000 kJ

25. At what temperature does the equilibrium constant for the formation of methane (CH_4) equal one when $\Delta H = -74.81$ kJ and $\Delta S = -80.8$ J/K?

a. 1080 K   b. 926 K   c. 762 K   d. 474 K   e. 1.08 K

26. The first law of thermodynamics states that

a. the energy of every pure substance is zero.
b. disorder is always increasing.
c. enthalpy is always increasing.
d. the total energy of the universe is constant.
e. the entropy of the surroundings is equal to zero.

27. The second law of thermodynamics states that

a. heat is energy.
b. the enthalpy of the universe in increasing.
c. ΔS of the universe is equal to zero.
d. if ΔG is negative, the reaction is spontaneous.
e. the total entropy of the universe is increasing.

28. Which of the following statements summarizes the third law of thermodynamics?

a. The entropy of every pure, perfectly crystalline substance at absolute zero is zero.
b. The energy of the universe is constant.
c. The entropy is always increasing.
d. When ΔH is negative, the reaction is spontaneous.
e. At higher temperatures, a reaction will always occur.

29. Calculate $\Delta G°$ for the following reaction:

$$CH_4(g) + 2\,O_2(g) \rightarrow CO_2(g) + 2H_2O(g)$$

given that $\Delta G°_f\,[CO_2(g)] = -394.4$ kJ/mol, $\Delta G°_f\,[CH_4(g)] = -50.7$ kJ/mol, and $\Delta G°_f\,[H_2O(g)] = -228.6$ kJ/mol.

a. -800.9 kJ   b. -673.7 kJ   c. -572.3 kJ   d. -436.4 kJ   e. 572.3 kJ

30. Calculate $\Delta G°$ for the following reaction:

$$2CO_2(g) \rightarrow 2CO(g) + O_2(g)$$

given that $\Delta G°_f\,[CO(g)] = -137.2$ kJ/mol and $\Delta G°_f\,[CO_2(g)] = -394.4$ kJ/mol.

a. -257.2 kJ   b. 257.2 kJ   c. 514.4 kJ   d. 531.6 kJ   e. 788.8 kJ

31. Given the following information, calculate $\Delta G°$ for the reaction below at 25°C:

$H_2O(g) + S(s) \rightarrow H_2S(g) + 1/2\ O_2(g)$

$\Delta H° = +221.2$ kJ and $\Delta S° = +87.7$ J/K

a. -219.0 kJ b. -195.1 kJ c. 195.1 kJ d. 219.0 kJ e. 247.3 kJ

32. Given the following information, calculate $\Delta G°$ for the reaction below at 25°C:

$SnCl_4(\ell) + 2H_2O(\ell) \rightarrow SnO_2(s) + 4HCl(g)$

$\Delta H° = 133.0$ kJ and $\Delta S° = 401.5$ J/K

a. -252.6 kJ b. -13.4 kJ c. 13.4 kJ d. 122.9 kJ e. 252.6 kJ

33. Given the following information, calculate $\Delta G°$ for the reaction below at 25°C:

$2H_2O_2(\ell) \rightarrow 2H_2O(\ell) + O_2(g)$

Compound	$\Delta H°_f$(kJ/mol)	S°(J/K·mol)
$H_2O_2(\ell)$	-187.8	109.6
$H_2O(\ell)$	-285.8	69.9
$O_2(g)$	———	205.1

a. -37700 kJ b. -342.6 kJ c. -233.5 kJ d. -192.3 kJ e. -157.9 kJ

34. Given the following information, calculate $\Delta G°$ for the reaction below at 25°C:

$2Al(s) + 3/2\ O_2(g) \rightarrow Al_2O_3(s)$

Compound	$\Delta H°_f$(kJ/mol)	S°(J/K·mol)
Al	———	28.3
O_2	———	205.1
Al_2O_3	-1676	50.92

a. -1768 kJ b. -1684 kJ c. -1668 kJ d. -1612 kJ e. -1582 kJ

35. A measure of the free energy of a reaction system is best given by its

1. change in enthalpy.
2. change in entropy.
3. change in heat capacity.

a. 1 only b. 2 only c. 3 only d. 1 and 2 only e. 1, 2, and 3

36. For the process at 25°C $I_2(g) \rightarrow I_2(s)$, what are the signs of ΔG, ΔH, and ΔS?

	ΔG	ΔH	ΔS
a.	+	-	-
b.	-	-	-
c.	-	+	+
d.	-	-	+
e.	+	+	+

37. The reaction of atomic oxygen with oxygen to give ozone is represented by the equation

$$O(g) + O_2(g) \rightarrow O_3(g)$$

The reaction is spontaneous at 25°C and 1 atm pressure. Under these conditions, we would expect that

a. $\Delta H°$ is negative and $\Delta G° < O$ and $K > 1$.

b. $\Delta H°$ is negative and $\Delta G° < O$ and $K < 1$.

c. $\Delta H°$ is positive and $\Delta G° > O$ and $K < 1$.

d. $\Delta H°$ is positive and $\Delta G° > O$ and $K > 1$.

e. $\Delta H°$ is zero and $\Delta G° = 0$ and $K = 1$.

38. For the reaction

$$MgO(s) + CO_2(g) \rightarrow MgCO_3(s)$$

$\Delta H°_{rxn} = -178$ kJ and $\Delta S°_{rxn} = -161$ J/mol·K.

Will the reaction be spontaneous at 900°C?

a. Yes, because ΔG will change.

b. Yes, because ΔH and ΔS are temperature independent.

c. Yes, because ΔH and ΔS are temperature dependent.

d. No, because ΔG is positive.

e. No, because ΔG is negative.

39. If a process is exothermic and not spontaneous, then what must be true?

a. $\Delta S > O$ b. $\Delta H > O$ c. $\Delta G = O$ d. $\Delta S < O$ e. $\Delta H = O$

40. If ΔS and ΔH are both positive for a given reaction and the reaction is not spontaneous at room temperature, which of the following must be true?

a. The reaction is spontaneous at high temperatures.

b. The reaction will never be spontaneous.

c. ΔH is the driving force.

d. The reaction is not spontaneous at higher temperatures.

e. Spontaneity does not depend on temperature.

41. For any reaction at equilibrium, which of the following is true?

a. $\Delta H<O$ b. $\Delta S=O$ c. $\Delta S<O$ d. $\Delta H=O$ e. $\Delta G=O$

42. Which of the following is true about vaporization?

a. ΔS is positive and ΔH is negative.

b. ΔS, ΔH, and ΔG are all negative.

c. ΔS and ΔH are both negative.

d. ΔS and ΔH are both positive.

e. ΔS, ΔH, and ΔG are equal to zero.

43. If a forward reaction has $\Delta G>O$, which of the following is true?

a. ΔS and ΔH are both positive.

b. ΔS and ΔH are negative.

c. ΔS is positive and ΔH is negative.

d. The reaction will not occur under any conditions.

e. The reverse reaction is spontaneous.

44. All of the following have $\Delta G°_f = O$ **EXCEPT**

a. $O_2(g)$ b. $Br_2(g)$ c. $H_2(g)$ d. Ca(s) e. Hg(ℓ)

45. Which of the following does not have a free energy of zero?

a. N(g) b. $I_2(s)$ c. Fe(s) d. He(g) e. He(g)

46. Ammonium nitrate spontaneously dissolves in water at room temperature and the process causes the solution to become quite cold. Which of the following is **TRUE** about the dissolution of ammonium nitrate?

a. The process is exothermic.

b. Its solubility will be greater in warmer water.

c. $\Delta S°$ for the reaction is negative.

d. All solutions of ammonium nitrate are supersaturated.

e. All solutions of ammonium nitrate are cold.

47. When sodium is added to water, the following reaction takes place:

$$Na(s) + H_2O(\ell) \rightarrow NaOH(aq) + 1/2\ H_2(g); \qquad \Delta H<O$$

What can be said about this reaction?

a. It is not possible.

b. ΔG will increase with increasing temperature.

c. The process will be non-spontaneous at low temperature.

d. The process will be non-spontaneous at 25°C.

e. ΔG will decrease with increasing temperature.

48. The following general reaction is not spontaneous at room temperature.

$$A + B \rightarrow C + D \qquad \Delta H° = +50.0\ kJ \text{ and } \Delta S° = +100.\ J/K$$

At what temperature will the reaction become spontaneous?

a. 500°C b. 0.5 K c. 500 K d. 250°C e. Not at any temperature.

49. In the reaction of two Cl atoms to give a Cl_2 molecule [2 Cl(g) → Cl_2(g)], the enthalpy change (ΔH) is __________ and the sign of the entropy change (ΔS) is __________.

a. negative, positive b. positive, negative c. negative, negative d. positive, positive

e. zero, positive

50. Which of the following would you expect to have the largest entropy?

a. He(g) b. $CaCO_3$(s) c. C_2H_6(g) d. CH_4(g) e. $HNO_3(\ell)$

Chapter 20: Answers:

1. a
2. c
3. b
4. b
5. c
6. a
7. d
8. d
9. b
10. d
11. c
12. a
13. b
14. e
15. c
16. a
17. a
18. a
19. a
20. b
21. c
22. a
23. b
24. a
25. b
26. d
27. e
28. a
29. a
30. c
31. c
32. c
33. c
34. e
35. d
36. b
37. a
38. d
39. d
40. a
41. e
42. d
43. e
44. b
45. a
46. b
47. e
48. c
49. c
50. c

Chapter 21
Principles of Reactivity: Electron Transfer Reactions

The following reduction potential table will either be necessary or helpful for questions 6, 7, 8, 13, 14, 15, 16, 17, 28, 29, 31, 43, 44.

Standard Reduction Potentials in Aqueous Solution at 25°C

			E° (volts)
$F_2(g)$ + 2e	→	$2F^-(aq)$	+2.87
$Au^{3+}(aq)$ + 3e	→	Au(s)	+1.50
$Cl_2(g)$ + 2e	→	$2Cl^-(aq)$	+1.36
$O_2(g)$ + $4H_3O^+(aq)$ + 4e	→	$6H_2O(\ell)$	+1.23
$Br_2(\ell)$ + 2e	→	$2Br^-(aq)$	+1.08
$Ag^+(aq)$ + e	→	Ag(s)	+0.80
$I_2(s)$ + 2e	→	$2I^-(aq)$	+0.535
$Cu^{2+}(aq)$ + 2e	→	Cu(s)	+0.337
$Sn^{4+}(aq)$ + 2e	→	$Sn^{2+}(aq)$	+0.15
$Sn^{2+}(aq)$ + 2e	→	Sn(s)	-0.14
$Cd^{2+}(aq)$ + 2e	→	Cd(s)	-0.40
$Zn^{2+}(aq)$ + 2e	→	Zn(s)	-0.763
$2H_2O(\ell)$ + 2e	→	$H_2(g)$ + $2OH^-(aq)$	-0.828
$Al^{3+}(aq)$ + 3e	→	Al(s)	-1.66
$K^+(aq)$ + e	→	K(s)	-2.93
$Li^+(aq)$ + e	→	Li(s)	-3.045

1. Which of the following is the correct cell notation for the reaction

$$Hg_2^{2+} + Cd(s) \rightarrow Cd^{2+} + 2Hg(\ell)$$

a. $Cd^{2+} \mid Cd \parallel Hg_2^{2+} \mid Hg$ b. $Cd^{2+} \mid Hg_2^{2+} \parallel Cd \mid Hg$ c. $Cd \mid Cd^{2+} \parallel Hg_2^{2+} \mid Hg$
d. $Cd^{2+} \mid Hg \parallel Hg_2^{2+} \mid Cd$ e. $Hg \mid Cd \parallel Hg_2^{2+} \mid Cd^{2+}$

2. Which of the following is the correct cell notation for the reaction

$$Au^{3+} + Al(s) \rightarrow Al^{3+} + Au(s)$$

a. $Al^{3+} \mid Al \parallel Au^{3+} \mid Au$ b. $Al \mid Al^{3+} \parallel Au^{3+} \mid Au$ c. $Al^{3+} \mid Au^{3+} \parallel Al \mid Au$
d. $Al^{3+} \mid Au \parallel Au^{3+} \mid Al$ e. $Au \mid Al \parallel Au^{3+} \mid Al^{3+}$

3. Consider an electrochemical cell where the following reaction takes place:

$$3Sn^{2+}(aq) + 2Al(s) \rightarrow 3Sn(s) + 2Al^{3+}(aq)$$

Which of the following is the correct cell notation for this cell?

a. $Al \mid Al^{3+} \parallel Sn^{2+} \mid Sn$ b. $Al^{3+} \mid Al \parallel Sn \mid Sn^{2+}$ c. $Sn \mid Sn^{2+} \parallel Al^{3+} \mid Al$
d. $Sn \mid Al^{3+} \parallel Al \mid Sn^{2+}$ e. $Al \mid Sn^{2+} \parallel Sn \mid Al^{3+}$

4. E° for the following redox reaction is -0.029 V:

$$Fe^{3+}(aq) + Ag(s) \rightarrow Fe^{2+}(aq) + Ag^{+}(aq)$$

What is ΔG° for this reaction?

a. +2.8 kJ   b. -2.8 kJ   c. +1.4 kJ   d. +8.4 kJ   e. -8.4 kJ

5. An early method of producing aluminum metal was the reaction of aluminum salts with sodium metal:

$$Al^{3+} + 3Na(s) \rightarrow Al(s) + 3Na^{+} \qquad E° = +1.05\ V$$

What is ΔG° for this reaction?

a. -304 kJ   b. -101 kJ   c. +101 kJ   d. +202 kJ   e. +304 kJ

6. Calculate ΔG for the following reaction:

$$2Ag^{+}(aq) + Sn(s) \rightarrow 2Ag(s) + Sn^{2+}(aq)$$

a. +64 kJ   b. +91 kJ   c. +181 kJ   d. -64 kJ   e. -181 kJ

7. Calculate ΔG for the following reaction:

$$I_2(s) + 2Br^{-}(aq) \rightarrow 2I^{-}(aq) + Br_2(\ell)$$

a. +105 kJ   b. -105 kJ   c. +312 kJ   d. +52 kJ   e. -312 kJ

8. Calculate ΔG for the following reaction:

$$2Au^{3+}(aq) + 3Zn(s) \rightarrow 2Au(s) + 3Zn^{2+}(aq)$$

a. +1310 kJ   b. +655 kJ   c. -437 kJ   d. -1310 kJ   e. -655 kJ

9. If ΔG of the following reaction is -203 kJ, what is E°?

$$2Ag^+(aq) + Ni(s) \rightarrow 2Ag(s) + Ni^{2+}(aq)$$

a. -1.05 V b. +2.10 V c. +0.0011 V d. -0.011 V e. +1.05 V

10. If ΔG of the following reaction is -144 kJ, what is E°?

$$A^{3+}(aq) + 3B(s) \rightarrow A(s) + 3B^+(aq)$$

a. +0.394 V b. +0.591 V c. +1.18 V d. -0.591 V e. -1.18 V

11. Given the two half reactions and their potentials, which net reaction is spontaneous?

$Mg^{2+}(aq) + 2e^- \rightarrow Mg(s)$ E° = -237 V

$Ni^{2+}(aq) + 2e^- \rightarrow Ni(s)$ E° = -0.25 V

a. $Ni(s) + Mg^{2+}(aq) \rightarrow Mg(s) + Ni^{2+}(aq)$ b. $Ni^{2+}(aq) + Mg(s) \rightarrow Mg^{2+}(aq) + Ni(s)$

c. $Ni(s) + Mg(s) \rightarrow Mg^{2+}(aq) + Ni^{2+}(aq)$ d. $Mg^{2+}(aq) + Ni^{2+}(aq) \rightarrow Mg(s) + Ni(s)$

e. $Mg^{2+}(aq) + Mg(s) \rightarrow Ni(s) + Ni^{2+}(aq)$

12. Given the two half-reactions and their potentials, which equation represents a spontaneous reaction?

$Zn^{2+}(aq) + 2e^- \rightarrow Zn(s)$

$Ni^{2+}(aq) + 2e^- \rightarrow Ni(s)$

a. $Ni^{2+}(aq) + Zn^{2+}(aq) \rightarrow Ni(s) + Zn(s)$ b. $Ni^{2+}(aq) + Zn(s) \rightarrow Ni(s) + Zn^{2+}(aq)$

c. $Ni(s) + Zn^{2+}(aq) \rightarrow Zn(s) + Ni^{2+}(aq)$ d. $Zn(s) + Ni(s) \rightarrow Ni^{2+}(aq) + Zn^{2+}(aq)$

e. $Zn(s) + Zn^{2+}(aq) \rightarrow Ni^{2+}(aq) + Ni(s)$

13. Calculate E° for the following reaction:

$$Sn^{4+}(aq) + 2K(s) \rightarrow Sn^{2+}(aq) + 2K^+(aq)$$

a. +6.00 V b. -3.08 V c. +3.08 V d. +2.78 V e. -2.78 V

14. Calculate E° for the following reaction:

$$2F^-(aq) + Cl_2(g) \rightarrow F_2(g) + 2Cl^-(aq)$$

a. -1.51 V b. +8.46 V c. -4.23 V d. -8.46 V e. +4.23 V

15. Calculate E° for the following reaction:

$$2Al^{3+}(aq) + 3Cd(s) \rightarrow 2Al(s) + 3Cd^{2+}(aq)$$

a. -2.06 V b. +4.52 V c. +2.06 V d. -4.52 V e. -1.26 V

16. Using data from the reduction potential table and the reaction

$$2Na(s) + F_2(g) \rightarrow 2F^-(aq) + 2Na^+(aq) \qquad E° = +5.58\ V$$

calculate the standard oxidation potential of the half-reaction

$$Na(s) \rightarrow Na^+(aq) + e^-$$

a. -1.36 V b. +8.45 V c. -2.71 V d. +2.71 V e. -8.45 V

17. Using data from the reduction potential table and the reaction

$$2Ag(s) + Pt^{2+}(aq) \rightarrow Pt(s) + 2Ag^+(aq) \qquad E° = 0.38\ V$$

calculate the standard reduction potential of the half-reaction

$$Pt^{2+}(aq) + 2e^- \rightarrow Pt(s)$$

a. -1.18 V b. -0.40 V c. 0.40 V d. 1.18 V e. 2.00 V

18. If an electrochemical cell with the notation $Rb|Rb^+||Na^+|Na$ has a standard potential of +0.23 V, what is the standard reduction potential of the Rb half cell if the Na^+/Na half-reaction has a reduction potential of -2.71 V?

a. +2.50 V b. -2.50 V c. -2.94 V d. +2.94 V e. 2.30 V

19. An electrochemical cell of notation $Pd|Pd^{2+}||Cu^{2+}|Cu$ has E° = -0.65 V. If we know that the standard reduction potential of Cu^{2+}/Cu is E° = 0.34 V, what is the standard reduction potential for Pd^{2+}/Pd?

a. -0.99 V b. -0.31 V c. +0.31 V d. 0.62 V e. +0.99 V

20. Given the following two half-reactions

$$Cd^{2+}(aq) + 2e^- \rightarrow Cd(s) \qquad E° = -0.40\ V$$

$$Zr^{4+}(aq) + 4e^- \rightarrow Zr(s) \qquad E° = -1.53\ V$$

determine E° and the spontaneity of the following reaction:

$$2Cd^{2+}(aq) + Zr(s) \rightarrow 2Cd(s) + Zr^{4+}(aq)$$

a. +1.13 V and not spontaneous b. +1.13 V and spontaneous c. -1.13 V and not spontaneous
d. -1.13 V and spontaneous e. -1.93 V and not spontaneous

21. What is the equilibrium constant for the following reaction at 298 K?

$$2Ag^+(aq) + 2I^-(aq) \rightarrow I_2(s) + 2Ag(s) \qquad E° = +0.265\ V$$

a. 2.99×10^4 b. 9.04×10^8 c. 7.73×10^3 d. 87.9 e. 1.60×10^7

22. What is the equilibrium constant for the following reaction at 37°C?

$$Hg_2^{2+}(aq) + 2Cl^-(aq) \rightarrow 2Hg(\ell) + Cl_2(g) \qquad E° = -0.57\ V$$

a. 5.1×10^{-20} b. 1.7×10^{-43} c. 2.1×10^{28} d. 2.9×10^{-19} e. 2.0×10^{19}

23. What is the equilibrium constant for the following reaction at 20°C?

$Fe(s) + Cu^{2+}(aq) \rightarrow Fe^{2+}(aq) + Cu(s)$ $E° = +0.78$ V

a. 2.3×10^{26} b. 6.9×10^{26} c. 1.4×10^{27} d. 1.8×10^{28} e. 1.2×10^{-21}

24. What is the cell potential for

$Fe(s) + Cd^{2+}(aq) \rightarrow Fe^{2+}(aq) + Cd(s)$ $E° = 0.040$ V

when $[Fe^{2+}] = 0.020$ and $[Cd^{2+}] = 0.20$ at 298 K?

a. +0.019 V b. + 0.039 V c. +0.010 V d. +0.099 V e. +0.070 V

25. What is the cell potential for

$3Sn^{4+}(aq) + 2Al(s) \rightarrow 3Sn^{2+}(aq) + 2Al^{3+}(aq)$ $E° = 1.81$ V

when $[Sn^{4+}] = 1.0$, $[Sn^{2+}] = 1.0 \times 10^{-2}$, and $[Al^{3+}] = 1.5 \times 10^{-3}$ at 298 K.

a. 1.70 V b. 1.76 V c. 1.81 V d. 1.86 V e. 1.93 V

26. The equilibrium constant for the following reaction is 1.65×10^{16} at 298 K. What is E°?

$MnO_2(s) + 4H^+(aq) + 2Cl^-(aq) + SbCl_4^-(aq) \rightarrow Mn^{2+}(aq) + SbCl_6^-(aq) + 2H_2O(\ell)$

a. +1.11 V b. +0.96 V c. +0.48 V d. -0.86 V e. -0.43 V

27. If the potential of a cell is +1.32 V at Q = 0.0969 with n = 2, what is the standard potential of the cell?

a. +1.35 V b. +1.48 V c. +1.31 V d. +1.34 V e. +1.29 V

28. Predict the product at the cathode when electric current is passed through a solution of KI.

a. K(s) b. $H_2(g)$ c. $I_2(\ell$ d. $O_2(g)$ e. $H_2O(\ell)$

29. Predict the product at the anode when electric current is passed through a solution of KI.

a. $I_2(\ell)$ b. $K^+(aq)$ c. $H_2(g)$ d. K(s) e. $O_2(g)$

30. If electric current is passed through molten $CuCl_2$, the product at the anode would be ________ and the product at the cathode would be ________.

a. $H_2O(\ell)$, $Cl_2(g)$ b. Cu(s), $Cl_2(g)$ c. $H_2(g)$, Cu(s) d. $Cl_2(g)$, Cu(s) e. Cu(s), $H_2O(\ell)$

31. If electric current is passed through aqueous LiBr, the product at the cathode would be _______ and the product at the anode would be ________.

a. $H_2O(\ell)$, $Li^+(aq)$ b. $Br_2(\ell)$, Li(s) c. Li(s), $Br_2(\ell)$ d. $Br_2(\ell)$, $H_2(g)$ e. $H_2(g)$, $Br_2(\ell)$

32. How many coulombs of charge are required to deposit 1.00 g Ag from a solution of $Ag^+(aq)$?

a. 9.27×10^{-3} b. 1.00 c. 894 d. 1230 e. 1790

33. How long would it take to deposit 1.36 g of copper from an aqueous solution of copper(II) sulfate by passing a current of two amperes through the solution?

a. 2070 sec b. 1.11 x 10^{-5} sec c. 2570 sec d. 736 sec e. 1030 sec

34. If you wish to plate 1.5 g of gold by electrolyzing a solution of Au^{3+} with 2.5 amperes, how long would it take?

a. 58,000 sec b. 880 sec c. 2.1 x 10^{-8} sec d. 500 sec e. 290 sec

35. If a current of 6.0 amps is passed through a solution of Ag^{+} for 1.5 hours, how many grams of silver are produced?

a. 0.60 g b. 36 g c. 0.34 g d. 3.0 g e. 1.0 g

36. How much platinum would be produced by passing a 2.0 ampere current through a solution of Pt^{2+} for 30. minutes?

a. 15 g b. 7.3 g c. 3.6 g d. 1.8 g e. 0.91 g

37. How many kilowatt hours of electrical energy are required to plate 2.00 grams of silver from an aqueous solution of silver nitrate on to a necklace using 3.00 V? (1 joule = 1 volt·coulomb and 1 kwh = 3.60 x 10^{6} J)

a. 0.00135 kwh b. 0.000165 kwh c. 32.4 kwh d. 0.00149 kwh e. 2.07 kwh

38. How many kwh of electrical energy are required to produce 1 kg of aluminum from a molten mixture of Al^{3+} using 2.5 V? (1 joule = 1 volt·coulomb and 1 kwh = 3.60 x 10^{6} J)

a. 7.5 kwh b. 2.5 kwh c. 0.40 kwh d. 1.2 kwh e. 3.8 x 10^{-4} kwh

39. How is aluminum currently produced in industry?

a. by reduction of Al^{3+} in Al_2O_3 with Na(s)
b. electrochemical reduction of pure Al_2O_3 to give Al and O_2
c. electrolysis of AlF_3 to give Al and F_2
d. electrolysis of a mixture of Al_2O_3 and Na_3AlF_6 to give Al and O_2
e. by reduction of Al^{3+} in Al_2O_3 with CO(g)

40. A possible anodic reaction that takes place during corrosion of iron is

a. $O_2(g) + 2H_2O(\ell) + 4e \rightarrow 4OH^-(aq)$. b. $Fe(s) \rightarrow Fe^{2+} + 2e$.
c. $2H_2O(\ell) + 2e \rightarrow H_2(g) + 2OH^-(aq)$. d. $SO_2(g) + O_2(g) + 2e \rightarrow SO_4^{2-}(aq)$ e. $2H^+ + 2e \rightarrow H_2(g)$

41. How was aluminum originally made?

a. the Hall-Heroult process

b. Al_2O_3 mixed with cryolite is electrolyzed

c. electrolysis of molten Al_2O_3

d. mining and purifying directly

e. reducing $AlCl_3$ with sodium

42. What metal is utilized at the anode of a mercury battery?

a. Pb b. Hg c. Zn d. Ni e. Pt

43. Using data from the reduction potential table, predict which of the following is the best oxidizing agent.

a. F_2 b. Ag c. Sn^{4+} d. Ag^+ e. Al^{3+}

44. Using data from the reduction potential table, predict which of the following is the best reducing agent.

a. Ag^+ b. Al c. F^- d. Sn^{2+} e. F_2

45. Under acidic conditions the bromate ion is reduced to the bromide ion. Write the balanced half-reaction for this process.

a. $BrO_3^- + 6H^+ + 6e \rightarrow Br^- + 3H_2O$ b. $2BrO_3^- + 6H^+ \rightarrow Br_2^- + 6H_2O + 3e$

c. $BrO_3^- + 6H_2O + 10\,e \rightarrow Br_2^- + 12H^+ + 3\,O_2$ d. $2BrO_3^- + 6H_2O \rightarrow 2Br^- + 12H^+ + 6\,O_2 + 8e$

e. $2BrO_3^- + 6H^+ \rightarrow Br_2^- + 3H_2O + 3e$

46. Balance the following redox equation which occurs in acidic solution.

$$Cu(s) + NO_3^-(aq) \rightarrow Cu^{2+}(aq) + NO(g)$$

a. $4H^+ + NO_3^- + Cu(s) \rightarrow Cu^{2+} + NO + 2H_2O$ b. $2H_2O + NO_3^- + Cu(s) \rightarrow NO + Cu^{2+} + 4H^+$

c. $2NO_3^- + 8H^+ + 3Cu(s) \rightarrow 3Cu^{2+} + 2NO + 4H_2O$ d. $3NO_3^- + 6H^+ + 2Cu(s) \rightarrow 2Cu^{2+} + NO + 3H_2O$

e. $6H^+ + 3NO_3^- + 2Cu^{2+} \rightarrow Cu(s) + NO + 3H_2O$

47. Balance the following redox equation which occurs in acidic solution.

$$N_2H_4(g) + BrO_3^-(aq) \rightarrow Br^-(aq) + N_2(g)$$

a. $3N_2H_4 + BrO_3^- \rightarrow 3N_2 + Br^- + 3H_2O + 6H^+$ b. $N_2H_4 + BrO_3^- + 2H^+ \rightarrow 2Br^- + N_2 + 3H_2O$

c. $3N_2H_4 + 2BrO_3^- + 12H^+ \rightarrow 3N_2 + 2Br^- + 6H_2O + 12H^+$

d. $N_2H_4 + 2BrO_3^- + 8H^+ \rightarrow 2Br^- + N_2 + 6H_2O$ e. $3N_2H_4 + 2BrO_3^- \rightarrow 3N_2 + 2Br^- + 6H_2O$

48. What is the balanced reduction half-reaction under acidic conditions in the equation below?

$$SO_2 + Cr_2O_7^{2-} \rightarrow Cr^{3+} + SO_4^{2-}$$

a. $14H^+ + Cr_2O_7^{2-} \rightarrow 2Cr^{3+} + 7H_2O + 8e$
b. $14H^+ + Cr_2O_7^{2-} \rightarrow 2Cr^{3+} + 7H_2O + 6e$
c. $8e + 14H^+ + Cr_2O_7^{2-} \rightarrow 2Cr^{3+} + 7H_2O$
d. $6e + 14H^+ + Cr_2O_7^{2-} \rightarrow 2Cr^{3+} + 7H_2O$
e. $9e + 14H^+ + Cr_2O_7^{2-} \rightarrow 2Cr^{3+} + 7H_2O$

49. Which of the following reactions is NOT a redox reaction?

a. $2HgO(s) \rightarrow 2Hg(\ell) + O_2(g)$
b. $H_2(g) + Br_2(g) \rightarrow 2HBr(g)$
c. $2HCl(aq) + Zn(s) \rightarrow H_2(g) + ZnCl_2(aq)$
d. $H_2CO_3(aq) \rightarrow H_2O(\ell) + CO_2(g)$
e. $2KClO_3 \rightarrow 2KCl(s) + 3\ O_2(g)$

50. Balance the following redox reaction which occurs in acidic solution:

$$Sn(s) + NO_3^-(aq) \rightarrow Sn^{4+}(aq) + N_2O(g)$$

a. $Sn(s) + 10H^+ + 2NO_3^- \rightarrow N_2O + Sn^{4+} + 5H_2O$
b. $2Sn(s) + 10H^+ + 2NO_3^- \rightarrow N_2O + 2Sn^{4+} + 5H_2O$
c. $3Sn(s) + 4H^+ + 2NO_3^- \rightarrow 2N_2O + 3Sn^{4+} + 2H_2O$
d. $2Sn(s) + 2H^+ + 2NO_3^- \rightarrow N_2O + 2Sn^{4+} + H_2O$
e. $Sn(s) + 6H^+ + 2NO_3^- \rightarrow N_2O + 2Sn^{4+} + 3H_2O$

Chapter 21: Answers:

1. c	26. c
2. b	27. e
3. a	28. b
4. a	29. a
5. a	30. d
6. e	31. e
7. a	32. c
8. d	33. a
9. e	34. b
10. a	35. b
11. b	36. c
12. b	37. d
13. c	38. a
14. a	39. d
15. e	40. b
16. d	41. e
17. d	42. c
18. c	43. a
19. e	44. b
20. b	45. a
21. b	46. c
22. d	47. e
23. b	48. d
24. e	49. d
25. e	50. b

Chapter 22
The Chemistry of the Main Group Elements

1. Of the elements N, P, As, Sb, and Bi, which one has the most metallic character?

   a. N   b. P   c. As   d. Sb   e. Bi

2. A group of metals which all melt below 200°C are

   a. alkali metals.   b. transition metals.   c. alkaline earth metals.   d. precious metals.
   e. actinide metals.

3. Which property of metals decreases as one moves down a group in the periodic chart?

   a. atomic radius   b. ionic radius   c. ionization energy   d. atomic mass   e. atomic number

4. The primary product of the reaction of sodium with pure oxygen is not the anticipated sodium oxide. It is __________ with the formula __________.

   a. sodium peroxide, Na_2O b. sodium peroxide, NaO_2 c. sodium superoxide, Na_2O_2
   d. sodium peroxide, Na_2O_2 e. sodium superoxide, Na_2O_2

5. Potassium superoxide can be used to produce oxygen under very controlled conditions according to the equation:

   a. $2KO(s) + CO_2(g) \rightarrow K_2CO_3(s) + 1/2\ O_2(g)$ b. $K_2O_2(s) + CO_2(g) \rightarrow K_2CO_3(s) + 1/2\ O_2(g)$
   c. $4KO_2(s) + 2CO_2(g) \rightarrow 2K_2CO_3(s) + 3\ O_2(g)$ d. $KO_3(s) + 2CO_2(g) \rightarrow 2K_2CO_3(s) + 7/2\ O_2(g)$
   e. $2KO_4(s) + 2CO_2(g) \rightarrow 2K_2CO_3(s) + 3\ O_2(g)$

6. Synthesis gas (syngas) which is produced from coal also yields the element

   a. oxygen.   b. carbon.   c. hydrogen.   d. helium.   e. sulfur.

7. Which of the following methods is useful in the laboratory preparation of hydrogen?

   a. metal + acid   b. carbonate + acid   c. acid + base   d. acid + alcohol   e. all of these

8. The largest use of hydrogen gas is for the industrial production of

   a. aspirin.   b. alcohol.   c. gasoline.   d. sodium.   e. ammonia.

9. The second most abundant element in the earth's crust is

   a. aluminum.   b. hydrogen.   c. silicon.   d. sulfur.   e. carbon.

10. Oxides of the alkaline earth family form

a. basic solutions. b. acidic solutions. c. gases with water. d. noble gas compounds.

e. soluble sulfides.

11. The equation for the Haber Process is

a. $N_2(g) + 3H_2(g) \rightleftharpoons 2NH_3(g)$ b. $CH_4(g) + 2O_2(g) \rightarrow CO_2(g) + 2H_2O(g)$

c. $4CS_2(g) + 4O_2(g) \rightleftharpoons 4CO_2(g) + S_8(s)$ d. $N_2H_4(g) + O_2(g) \rightarrow N_2(g) + 2H_2O(g)$

e. $2NH_3(aq) + NaOCl(aq) \rightarrow N_2H_4(aq) + NaCl(aq) + H_2O(\ell)$

12. Oxides of nitrogen are known which have the following positive oxidation numbers of nitrogen.

a. +2, +4 b. +2, +4, +6 c. +1, +3, +5 d. +2, +4, +5 e. +1, +2, +3, +4, +5

13. Which gases are commercially obtained from the liquefaction of air?

a. nitrogen b. oxygen c. nitrogen and oxygen d. helium and nitrogen

e. hydrogen and oxygen

14. The Ostwald process is useful for the preparation of

a. ammonia from nitrogen and hydrogen. b. sulfur from iron sulfide. c. nitric acid from ammonia.

d. oxygen from sand. e. lead from lead sulfide.

15. The building block of the silicate minerals is the

a. tetrahedral SiO_4 unit. b. the bent SiO_2 unit. c. the bent SiO_4 unit.

d. the trigonal planar SiO_3 unit. e. the octahedral SiO_4 unit.

16. Which of the following metals is NOT attacked by nitric acid?

a. Fe b. Ti c. Au d. Cu e. Co

17. Aluminum is resistant to corrosion because

a. aluminum is a noble metal.

b. the surface of aluminum is coated with unreactive Al_2O_3.

c. aluminum will not gain electrons.

d. aluminum forms slightly acidic hydrates.

e. aluminum usually contains a trace amounts of other metals such as Cr^{3+}.

18. Which compound ranks #1 in terms of pounds produced annually in the United States?

a. ethanol b. ammonia c. benzene d. sulfuric acid e. sodium hydroxide

19. All of the following statements about the main group metals are true **EXCEPT**

a. most metal oxides are basic.

b. the metals have positive reduction potentials.

c. most metals are dense solids at 400 K.

d. the metals are good heat conductors.

e. the metals are good electrical conductors.

20. In which reaction is $H_2PO_4^-$ acting as a base?

a. $H_2PO_4^- + CN^- \rightleftharpoons HCN + HPO_4^{2-}$

b. $H_2PO_4^- + OH^- \rightleftharpoons H_2O + HPO_4^{2-}$

c. $H_2PO_4^- + HS^- \rightleftharpoons H_2S + HPO_4^{2-}$

d. $H_2PO_4^- + HF \rightleftharpoons F^- + H_3PO_4$

e. $H_2PO_4^- + NH_3 \rightleftharpoons NH_4^+ + HPO_4^{2-}$

21. All of the following are acid-base conjugate pairs **EXCEPT**

a. H_3O^+, OH^- b. H_2O, OH^- c. NH_4^+, NH_3 d. CH_3COOH, CH_3COO^- e. HPO_4^{2-}, PO_4^{3-}

22. $HF + H_2O \rightleftharpoons H_3O^+ + F^-$

The conjugate base of HF is

a. HF b. H_2F^+ c. F^- d. H_2O e. OH^-

23. The conjugate base of HNO_2 is

a. HNO_2 b. H_3O^+ c. H^+ d. NO_2^- e. H_2O

24. The hydronium ion concentration of a 0.00100 acetic acid solution is 1.34×10^{-4} M. The pH of the solution is

a. 3.00. b. 3.40. c. 3.87. d. 4.00. e. 4.13.

25. Which of the following species is the best reducing agent?

a. Cl_2 b. F_2 c. Na d. Br^- e. O^{2-}

26. All of the following would be expected to function as oxidizing agents **EXCEPT**

a. Al^{3+}. b. Mg^{2+}. c. N^{3-}. d. NO_3^-. e. OCl^-.

27. All of the following would be expected to function as reducing agents **EXCEPT**

a. H_2. b. NH_3. c. Sn^{2+}. d. Mg. e. Al^{3+}.

28. All of the following ions can exist in aqueous solution **EXCEPT**

a. NH_4^+. b. Ca^+. c. K^+. d. Al^{3+}. e. Ba^{2+}.

29. All of the following reactions could be used to produce a protonic acid **EXCEPT**

a. $SO_2 + H_2O \rightarrow$ b. $SO_3 + H_2O \rightarrow$ c. $N_2O_5 + H_2O \rightarrow$ d. $CO_2 + H_2O \rightarrow$ e. $Na_2O + H_2O \rightarrow$

30. Tin has the oxidation states of +2 and +4. The expected oxidation states for antimony are

a. +3 and +5. b. +1 and +3. c. +2 and +3. d. +2 and +4. e. +1 and +5.

31. Hydrofluoric acid has a dissociation constant at 25°C of 6.7×10^{-4}. If 1.0 mole of HF is dissolved in enough water to give 1.00 L of solution, what is the hydrogen ion concentration in this solution expressed in moles/liter?

a. 2.0×10^{-2} to 2.4×10^{-2} b. 2.5×10^{-2} to 2.7×10^{-2} c. 3.0×10^{-3} to 3.5×10^{-3}

d. 6.6×10^{-4} to 6.8×10^{-4} e. None of these.

32. Calculate the dissociation constant of an acid if a 1.00 M solution of the acid has a pH of 3.18.

a. 4.4×10^{-7} b. 4.4×10^{-6} c. 6.6×10^{-4} d. 6.6×10^{-3} e. 3.2×10^{-3}

33. Which of the following equations is (are) balanced?

1. $NaBr(aq) + Pb(NO_3)_2(aq) \rightarrow PbBr_2(s) + NaNO_3(aq)$
2. $(NH_4)_2Cr_2O_7(s) \rightarrow N_2(g) + Cr_2O_3(s) + 4H_2O(\ell)$
3. $Fe_3O_4(s) + 3CO(g) \rightarrow 3Fe(s) + 3CO_2(g)$

a. 1 only b. 2 only c. 3 only d. 1 and 2 only e. 1, 2, and 3

34. $\_\_H_2SnCl_6 + \_\_H_2S \rightarrow \_\_SnS_2 + \_\_HCl$

When the equation above is properly balanced with the smallest whole numbers, the respective coefficients are:

a. 1, 2, 1, 6 b. 1, 1, 1, 3 c. 2, 4, 2, 6 d. 2, 4, 2, 12 e. 1, 4, 2, 6

35. Balance the following equation with the **smallest whole number coefficients** possible. Select the number that is the sum of the coefficients in the balanced equation:

$$\_\_KClO_3 \rightarrow \_\_KCl + \_\_O_2$$

a. 5 b. 6 c. 7 d. 8 e. 9

36. When the expression

$$C_5H_6NS(\ell) + O_2(g) \rightarrow CO_2(g) + H_2O + NO_2(g) + SO_2(g)$$

is balanced, the sum of all the smallest whole number coefficients is

a. 28 b. 29 c. 33 d. 36 e. 39

37. The complete combustion of butanone yields carbon dioxide and water. Balance the equation with the smallest whole number coefficients possible. Select the number that is the sum of the coefficients in the balanced equation.

$$\_\_CH_3CH_2COCH_3(\ell) + \_\_O_2(g) \rightarrow \_\_CO_2(g) + \_\_H_2O(\ell)$$

a. 15 b. 17 c. 27 d. 29 e. 30

38. $\_\_Nb_2O_5 + \_\_CCl_4 \rightarrow \_\_NbCl_5 + \_\_COCl_2$

a. 12 b. 13 c. 16 d. 18 e. 21

39. Choose the compound that is the most soluble in water.

a. HgS b. CuS c. Ag_2S d. FeS e. K_2S

40. A strong acid and weak acid respectively are

a. HF and HCl. b. H_2SO_4 and H_2SO_3. c. H_3PO_3 and H_3PO_4. d. HCl and HBr.
e. CH_3COOH and CH_3CH_2COOH.

41. All of the following are true statements about bases **EXCEPT**

a. they have a bitter taste.
b. they react with salts to form weaker or more volatile acids and a new salt.
c. they have a slippery feeling.
d. they change the colors of many indicators.
e. they react with acids to form salts and water.

42. Which of the following changes requires a reducing agent?

a. $Cr_2O_7^{2-} \rightarrow CrO_4^{2-}$ b. $ClO_4^- \rightarrow ClO_3^-$ c. $NH_3 \rightarrow NH_4^+$ d. $Al \rightarrow Al^{3+}$ e. $O^{2-} \rightarrow O_2^{2-}$

43. The following reaction is used for the analytical determination of uranium in **acidic** solution:

$$UO^{2+} + Cr_2O_7^{2-} \rightarrow UO_2^{2+} + Cr^{3+} \quad \text{(unbalanced)}$$

For every mole of UO^{2+} in an unknown _____ mole(s) of $Cr_2O_7^{2-}$ would be required for the above reaction.

a. 1/3 b. 1/2 c. 2 d. 3 e. 6

44. Balance the equation below. Choose from the numbers given as answers the one which is the coefficient of the underlined substance in the balanced ionic reaction:

$$\underline{Zn} + NO_3^- \rightarrow Zn(OH)_4^{2-} + NH_3 \quad \text{(basic solution)}$$

a. 1 b. 2 c. 4 d. 6 e. 8

45. In an electrolytic cell, the electrode that acts as a source of electrons to the solution is called the _______________. The chemical change that occurs at this electrode is called _______________.

a. cathode, reduction b. cathode, oxidation c. anode, oxidation d. anode, reduction

e. Cannot tell unless we know the species that are involved in the cell reaction.

46.

Half Reaction	E° (volts)
$Mn^{2+} + 2e \rightarrow Mn$	-1.03
$Sn^{2+} + 2e \rightarrow Sn$	-0.14
$Hg_2Cl_2 + 2e \rightarrow 2Hg + 2Cl^-$	0.27
$Cu^+ + e \rightarrow Cu$	0.52
$Ag^+ + e \rightarrow Ag$	0.80
$Br_2 + 2e \rightarrow 2Br^-$	1.09

The strongest reducing agent of this series is

a. Mn^{2+} b. Br_2 c. Br^- d. Sn e. Mn

47.

Half Reaction	E° (volts)
$Br_2 + 2e \rightarrow 2Br^-$	1.09
$Hg_2^{2+} + 2e \rightarrow 2Hg$	0.80
$Cu^{2+} + 2e \rightarrow Cu$	0.34
$2H^+ + 2e \rightarrow H_2$	0.00
$Sn^{2+} + 2e \rightarrow Sn$	-0.14
$Fe^{2+} + 2e \rightarrow Fe$	-0.41
$Al^{3+} + 3e \rightarrow Al$	-1.67

The strongest reducing agent of this series is

a. H_2 b. Br_2 c. Br^- d. Al e. Al^{3+}

48. Calculate the standard Gibbs free energy change for the reaction:

$$2Fe + 3Cl_2 \rightarrow 2Fe^{3+} + 6Cl^-$$

given the following electrode potentials:

	E°, volts
$Fe^{3+} + 3e \rightarrow Fe$	-0.036
$Cl_2 + 2e \rightarrow 2Cl^-$	1.358

a. 765 kJ b. -269 kJ c. -404 kJ d. -765 kJ e. -807 kJ

49. A cell consists of a magnesium electrode immersed in a solution of magnesium chloride and a silver electrode immersed in a solution of silver nitrate. The two half cells are connected by means of a salt bridge. It is possible to increase the voltage of the cell by

a. addition of sodium chloride to both half cells.

b. increasing the Mg^{2+} concentration and decreasing the Ag^{+} concentration.

c. increasing the size of the Mg electrode and decreasing the size of the Ag electrode.

d. decreasing the size of the Mg electrode and increasing the size of the Ag electrode.

e. decreasing the Mg^{2+} concentration.

50. Which of the following is obtained commercially by electrolytic **oxidation**?

a. bromine b. chlorine c. ammonia d. aluminum e. potassium iodide

Chapter 22: Answers:

1. e	26. c
2. a	27. e
3. c	28. b
4. d	29. e
5. c	30. a
6. c	31. b
7. a	32. a
8. e	33. b
9. c	34. a
10. a	35. c
11. a	36. e
12. e	37. d
13. c	38. b
14. c	39. e
15. a	40. b
16. c	41. b
17. b	42. b
18. d	43. a
19. b	44. c
20. d	45. a
21. a	46. e
22. c	47. d
23. d	48. e
24. c	49. e
25. c	50. b

Chapter 23
The Transition Elements

1. Which of the following groups of elements occur in nature in the free state?

   a. Na, K, Li b. Fe, Co, Mn c. Sc, Ti, V d. Pb, Bi, Sn e. Au, Pt, Ir

2. Which of the following transition metal compounds is used as a white paint pigment?

   a. CoO b. $CuSO_4$ c. NiO d. Cr_2O_3 e. TiO_2

3. In a blast furnace, iron oxides are reduced by

   a. $CaCO_3$ b. CO c. CO_2 d. $CaSO_4$ e. CS_2

4. Which of the following metals is most dense?

   a. Cr b. Ta c. W d. Hg e. Mo

5. Which element would have the highest melting point?

   a. Cd b. Ru c. La d. Hg e. Os

6. Pig iron is

   a. a soft and brittle form of iron that contains impurities.

   b. a hard and ductile form of iron that contains impurities.

   c. a hard very pure form of iron.

   d. a soft very pure form of iron.

   e. a partially oxidized form of iron.

7. Of the ions, Cr^{2+}, Zn^{2+}, and Ni^{2+}, which is (are) diamagnetic?

   a. Zn^{2+} only b. Cr^{2+} only c. Ni^{2+} only d. Zn^{2+} and Cr^{2+} e. all three ions

8. Copper ores are enriched to increase the percentage of copper in the mixture by a process called

   a. roasting. b. flotation. c. pyrometallurgy. d. hydrometallurgy. e. ganguation.

9. What is the oxidation number of the metal ion in $[Co(NH_3)_5SO_4]Cl$?

   a. +1 b. +2 c. +3 d. +5 e. +6

10. What is the oxidation number of the metal ion in $[Pt(NH_3)_2Cl_2]$?

    a. 0 b. +2 c. +4 d. +6 e. -1

11. An example of a neutral bidentate ligand is

a. ammonia. b. oxalate ($C_2O_4^{2-}$). c. acetate. d. ethylenediamine e. EDTA

12. The chelating agent EDTA is classified as a

a. neutral bidentate ligand.

b. neutral tetradentate ligand.

c. tetradentate ligand with a 2-charge.

d. tetradentate ligand with a 6-charge.

e. hexadentate ligand with a 4-charge.

13. The name of the coordination compound with the formula $(NH_4)_2[CuCl_4]$ is

a. ammonium tetrachlorocuprate(II).

b. diammonium copper(II) tetrachloride.

c. ammonium copper(II) chloride.

d. diammonium tetrachlorocopper(II).

e. copper(II) diamminetetrachloride.

14. The name of the coordination compound with the formula $Na[FeCl_4]$ is

a. sodium iron(III) tetrachloride.

b. sodium tetrachloroferide(III).

c. sodium chloroferrate(IV).

d. sodium tetrachloroferrate(III).

e. sodium ferroyltetrachloride.

15. The name of the coordination compound with the formula $[Co(en)_3]Cl_3$ is

a. (ethylenediamine) cobalt(III) chloride.

b. tris(ethylenediamine) cobalt(III) chloride.

c. ethylenediaminecobalt trichloride.

d. tris(ethylenediamine) cobalt trichloride.

e. (ethylenediamine) cobalt(IV) chloride.

16. The formula for triamminedichloronitritocobalt(III) is

a. $[Co(NH_3)ClNO_2]$. b. $[Co(NH_3)_3ClNO_2]$. c. $[Co(NH_3)_3Cl_2NO_2]$. d. $[Co(NH_3)_3(Cl_2)_2NO_2]$.
e. $[Co(NH_3Cl_2NO_2)_3]$.

17. The formula for the hydroxopentaaquairon(III) ion is

a. $[Fe(OH)(H_2O)_5]^{3+}$. b. $[Fe(OH)(H_2O)_5]^{2+}$. c. $[Fe(OH)_5]^{3+}(aq)$. d. $[(H_2O)_5Fe](OH)_3$.
e. $[Fe{\cdot}5\ H_2O](OH)_3$.

18. Which of the following can form geometric isomers?

a. $[Co(NH_3)_6]Cl_3$ b. $[Co(NH_3)_5Cl]^{2+}$ c. $[Co(NH_3)_4Cl_2]^+$ d. $[Co(NH_3)_5Cl]SO_4$
e. $[CoCl_4]^{2-}$

19. Which of the following can form optical isomers?

a. $CHCL_3$ b. CH_2Cl_2 c. $(CH_3)(CH_2CH_3)_2NH^+$ d. $BrCH(CH_3)CO_2H$ e. $CH_2(CO_2H)_2$

20. How many unpaired electrons are present in the high spin complex $[Fe(H_2O)_6]^{2+}$?

a. 8 b. 6 c. 4 d. 2 e. 0

21. Which of the following has a d^5 electron configuration?

a. $Co(CN)_4^-$ b. $Co(NH_3)_6^{3+}$ c. $RhCl_6^{4-}$ d. $V(CN)_6^{4-}$ e. $Fe(CN)_6^{3-}$

22. Which of the following involves the concentration of a metal ore?

1. flotation
2. roasting
3. filtration

a. 1 only b. 2 only c. 3 only d. 1 and 2 only e. 1, 2, and 3

23. Which of the following are important reducing agents in the recovery of metals from their ores?

1. C
2. CO
3. $CaSiO_3$

a. 1 only b. 2 only c. 3 only d. 1 and 2 only e. 1, 2, and 3

24. Which of the following occur as native ores?

1. Al
2. K
3. Au

a. 1 only b. 2 only c. 3 only d. 1 and 2 only e. 1, 2, and 3

25. All the properties listed below are characteristic of many of the transition element compounds **EXCEPT**

a. most of them are paramagnetic.

b. most of them are colored.

c. most of the elements upon ionizing lose the d electrons first.

d. most of the metals exhibit multiple oxidation states.

e. most of the elements form many complexes.

26. Which of the elements indicated below would be classed as transition elements?

a. $1s^2\ 2s^2\ 2p^5$ b. $1s^2\ 2s^2\ 2p^6\ 3s^2\ 3p^6$ c. $1s^2\ 2s^2\ 2p^6\ 3s^2\ 3p^6\ 3d^2\ 4s^2$

d. $1s^2\ 2s^2\ 2p^6\ 3s^2\ 3p^6\ 3d^{10}\ 4s^2\ 4p^1$ e. $1s^2\ 2s^2\ 2p^6\ 3s^2\ 3p^6\ 3d^{10}\ 4s^2\ 4p^6$

27. Transition metals can be distinguished from mail group metals by the fact that

a. main group metals only have +1 and +2 oxidation states.

b. main group metals have higher relative atomic masses than transition metals.

c. transition metals have higher relative atomic masses than the main group metals.

d. only the main group metals can form complex ions.

e. transition metals have a greater tendency to form colored compounds than main group metals.

28. All of the oxides given below would be expected to be able to react with more oxygen **EXCEPT**

a. Ti_2O_3. b. V_2O_3. c. Mn_2O_3. d. Sc_2O_3. e. Cr_2O_3.

29. What is the electron configuration of iron?

a. [Ar] $3d^6$ b. [Ar] $3d^6\ 4s^2$ c. [Ar] $3d^7\ 4s^1$ d. [Ar] $3d^8$ e. [Ar] $3d^5\ 4s^1$

30. What is the ground-state electronic configuration of Cr^{3+}?

a. [Ar] $4s^1\ 3d^5$ b. [Ar] $4s^2\ 3d^2$ c. [Ar] $4s^2\ 3d^1$ d. [Ar] $4s^1\ 3d^2$ e. [Ar] $3d^3$

31. What is the maximum oxidation state expected for vanadium?

a. +1 b. +2 c. +3 d. +4 e. +5

32. All of the following first transition series elements have a +4 oxidation state **EXCEPT**

a. Mn b. Cr c. V d. Sc e. Ti

33. All of the following elements from the first transition series and the preceding *s*-block elements reacts with hydrochloric acid to produce hydrogen **EXCEPT**

a. Ca b. Cr c. Mn d. Cu e. Zn

34. All of the following are characteristic of the transition metals **EXCEPT**

a. partially filled d or f orbitals.

b. variable oxidation states.

c. high densities.

d. the ability to form mainly diamagnetic complex compounds.

e. the ability to form mainly colored compounds.

35. Copper can be oxidized by

a. $CH_3COOH(aq)$. b. $HBr(aq)$. c. $HNO_3(aq)$. d. $H_2CO_3(aq)$. e. $H_2O(\ell)$.

36. The formula for a platinum(IV) complex is $[Pt(NH_3)_4Br_2]Cl_2$. In aqueous solution, it will dissociate into

a. 2 ions. b. 3 ions. c. 4 ions. d. 5 ions. e. 6 ions.

37. All of the following can function both as Lewis bases and as ligands **EXCEPT**

a. Cl^- b. OH^- c. NH_3 d. H_2O e. H_3O^+

38. A ligand is always

a. mondentate. b. a Lewis base. c. an ion. d. a molecule. e. paramagnetic.

39. How many isomers are possible for the square planar complex, $[Pt(NH_3)_2Br_2]$?

a. 1 b. 2 c. 3 d. 4 e. 6

40. How many isomers are possible for the square planar complex, $[Pt(NH_3)_3Br]Br$?

a. 1 b. 2 c. 3 d. 4 e. 6

41. The octahedral complex $[Fe(H_2O_6]^{2+}$ is a high spin complex. How many unpaired electrons does the iron ion have?

a. 1 b. 2 c. 3 d. 4 e. 5

42. How many unpaired electrons are there in the strong field complex, $[Co(NH_3)_6]^{3+}$?

a. 0 b. 1 c. 2 d. 3 e. 5

43. Which two elements can usually exhibit multiple oxidation states?

a. Na and Cl b. Mg and Se c. Al and Sc d. Cu and V e. Zn and Co

44. Considering only electron configurations of the metal ion, we would expect all of the following to be colored **EXCEPT**

a. $CrCl_3$ b. $ScCl_3$ c. $TiCl_3$ d. $CoCl_3$ e. $CuCl_2$

45. Considering only electron configurations of the metal ion, we would expect all the following to be colored **EXCEPT**

a. CuCl b. VCl_3 c. MnO_2 d. $CrCl_2$ e. $NiCl_2$

46. In which compound is cobalt in the highest oxidation state?

a. $K_4[Co(CN)_6]$ b. $Co_2(CO)_8$ c. $[Co(H_2O)_6]Cl_2$ d. $[Co(NH_3)_4Br_2]Cl$ e. $Na_2[CoCl_4]$

47. Which of the following Lewis bases would be expected to form chelates with transition metal ions?

1. $^{-}OOCCOO^{-}$
2. $(CH_3)_2NH$
3. EDTA

a. 1 only b. 2 only c. 3 only d. 1 and 3 only e. 1, 2, and 3

48. The total number of isomers possible for the octahedral complex $[Co(NH_3)_3Cl_3]$ is

a. 1 b. 2 c. 3 d. 4 e. 5

49. How many optical isomers do the cis and trans isomers of $[Co(en)_2Cl_2]^+$ have?

a. cis, 0 and trans, 2 b. cis, 2 and trans, 0 c. cis, 2 and trans, 2 d. cis, 0 and trans, 0
e. cis, 0 and trans, 1

50. The complex $[Co(NH_3)_6]Br_3$ is diamagnetic. Therefore, which set of terms best describes the complex

a. high spin, weak field, octahedral.
b. low spin, strong field, square planar.
c. low spin, strong field, octahedral.
d. high spin, strong field, octahedral.
e. low spin, weak field, square planar.

Chapter 23: Answers:

1. e	26. c
2. e	27. e
3. b	28. d
4. c	29. b
5. e	30. e
6. a	31. e
7. a	32. d
8. b	33. d
9. c	34. d
10. b	35. c
11. d	36. b
12. e	37. e
13. a	38. b
14. d	39. b
15. b	40. a
16. c	41. d
17. b	42. a
18. c	43. d
19. d	44. b
20. c	45. a
21. e	46. d
22. a	47. d
23. d	48. b
24. c	49. b
25. c	50. c

Chapter 24
Nuclear Chemistry

1. Which of the following symbols is used to represent gamma ray emissions?

   a. ${}^{-1}_{1}\gamma$ b. ${}^{0}_{0}\gamma$ c. ${}^{4}_{2}He$ d. ${}^{2}_{4}He$ e. ${}^{+1}_{-1}\gamma$

2. Which of the following symbols is used to represent alpha particle emissions?

   a. ${}^{-1}_{+1}\gamma$ b. ${}^{0}_{0}\gamma$ c. ${}^{4}_{2}He$ d. ${}^{2}_{4}He$ e. ${}^{+1}_{-1}\gamma$

3. Which of the following symbols represents positron emissions?

   a. ${}^{4}_{2}He^{2+}$ b. ${}^{2}_{4}He$ c. ${}^{0}_{-1}e$ d. ${}^{0}_{+1}e$ e. ${}^{0}_{0}e$

4. Which type of radiation is effectively blocked by a piece of paper?

   a. alpha particles b. beta particles c. electron capture d. positron emissions
   e. gamma radiation

5. Which of the following types of radiation has the greatest ability to penetrate matter?

   a. alpha rays b. beta rays c. gamma rays d. protons e. positrons

6. The atomic number of a nucleus which emits a beta particle will

   a. remain the same. b. increase by one unit. c. increase by two units.
   d. decrease by one unit. e. decrease by two units.

7. The atomic number of nucleus which emits a positron will

   a. remain the same. b. increase by one unit. c. increase by two units.
   d. decrease by one unit. e. decrease by two units.

8. The mass number of a nucleus which emits a positron will

   a. remain the same. b. increase by one unit. c. increase by two units.
   d. decrease by one unit. e. decrease by two units.

9. The atomic number of a nucleus which undergoes electron capture will

   a. remain the same. b. increase by one unit. c. increase by two units.
   d. decrease by one unit. e. decrease by two units.

10. If a nucleus decays by successive α, α, β decay, the atomic number will

a. increase by four units. b. increase by three units. c. increase by 1 unit.

d. decrease by eight units. e. decrease by three units.

11. If a nucleus decays by successive α, β, β emissions, how would the atomic number and mass number change?

a. The atomic number decreases by four; the mass number stays the same.

b. The atomic number increases by two; the mass number decreases by two units.

c. The atomic number stays the same; the mass number decreases by two units.

d. The atomic number decreases by two; the mass number decreases by four units.

e. The atomic number stays the same; the mass number decreases by four units.

12. When uranium-238 undergoes successive α, β, β, α emissions, the nucleus produced is

a. ^{84}Po b. ^{230}Th c. ^{227}U d. ^{230}Ra e. ^{228}Pa

13. When lead-210 undergoes successive β, β. α emissions, the nucleus produced is

a. ^{206}Pb b. ^{214}Rn c. ^{210}Po d. ^{204}Tl e. ^{228}Pa

14. The isotope $^{53}_{24}Cr$ is produced by the beta decay of

a. $^{53}_{25}Mn$ b. $^{54}_{24}Cr$ c. $^{52}_{24}Cr$ d. $^{53}_{23}V$ e. $^{54}_{23}V$

15. Elements below the peninsula of stability may

a. attain stability by beta emission because they have too few neutrons.

b. attain stability by beta emission because they have too many neutrons.

c. attain stability by positron emission because they have too many neutrons.

d. attain stability by positron emission because they have too few neutrons.

e. attain stability by electron capture because they have too many neutrons.

16. Complete the following fission reaction:

$$^{235}_{92}U + ^{1}_{0}n \rightarrow ^{139}_{53}I + \underline{\qquad} + 2\ ^{1}_{0}n$$

a. $^{95}_{39}Y$ b. $^{96}_{39}Y$ c. $^{96}_{40}Zr$ d. $^{95}_{40}Zr$ e. $^{94}_{41}Nb$

17. Complete the following nuclear reaction:

$$^{238}_{92}U + ^{12}_{6}C \rightarrow \_\_\_\_ + 6\,^{1}_{0}n$$

a. $^{249}_{99}Es$ b. $^{249}_{98}Cf$ c. $^{244}_{92}U$ d. $^{244}_{98}Cf$ e. $^{250}_{104}Rf$

18. What will complete the following equation?

$$^{9}_{4}Be + ^{4}_{2}He \rightarrow + \_\_\_\_ + ^{1}_{0}n$$

a. $^{13}_{7}N$ b. $^{14}_{7}N$ c. $^{12}_{6}C$ d. $^{13}_{6}C$ e. $^{14}_{6}C$

19. What particles are produced in the following reaction?

$$^{238}_{92}U + ^{16}_{8}O \rightarrow ^{250}_{100}Fm + \_\_\_\_$$

a. 2 neutrons b. 4 neutrons c. 1 alpha particle d. 2 alpha particles e. 4 alpha particles

20. All isotopes having an atomic number greater than that of the element ________ are radioactive.

a. lead b. bismuth c. strontium d. radium e. uranium

21. Which of the following isotopes is stable, despite having both an odd number of protons and also an odd number of neutrons?

a. $^{8}_{3}Li$ b. $^{14}_{6}C$ c. $^{24}_{11}Na$ d. $^{22}_{13}Al$ e. $^{14}_{7}N$

22. The point of maximum stability in the binding energy curve occurs in the vicinity of which of the following isotopes?

a. $^{4}_{2}He$ b. $^{56}_{26}Fe$ c. $^{150}_{62}Sm$ d. $^{207}_{82}Pb$ e. $^{235}_{92}U$

23. Which of the following elements undergoes nuclear fusion reactions to provide the primary source of energy from the sun?

a. uranium b. plutonium c. hydrogen d. iron e. helium

24. Which of the following particles causes a nuclear fission reaction in a uranium nucleus?

a. $^{0}_{-1}e$ b. $^{1}_{1}H$ c. $^{1}_{0}n$ d. $^{0}_{0}\gamma$ e. $^{0}_{+1}e$

25. What name is given to the amount of energy required to separate a nucleus into its individual nucleons (or the energy evolved when these nucleons combine to form a nucleus)?

a. nuclear fission b. nuclear fusion c. electron capture d. binding energy

e. ionization energy

26. What do scientists call the sequence of rapidly occurring reactions that results when a nuclear fission reaction produces enough neutrons to produce more fission reactions?

a. chain reaction b. nuclear fusion c. electron capture d. binding energy e. critical mass

27. The atoms $^{13}_{6}C$, $^{19}_{9}F$, and $^{23}_{11}Na$ have 13, 19, and 23 nucleons, respectively.

The total number of nucleons in the nucleus is equal to

a. the atomic number.

b. the number of psoitrons in the nucleus.

c. the number of protons in the nucleus.

d. the number of neutrons in the nucleus.

e. the number of protons and neutrons in the nucleus.

28. Beta emission can give increased nuclear stability by

a. increasing the mass ratio. b. decreasing the mass ratio. c. keeping the same mass ratio.

d. decreasing the n/p ratio. e. increasing the n/p ratio.

29. Positron emission can give increased nuclear stability by

a. keeping the same n/p ratio. b. decreasing the n/p ratio. c. increasing the n/p ratio.

d. decreasing the mass ratio. e. increasing the mass ratio.

30. The most unstable isotopes usually have ____ number of neutrons, ____ number of protons, and ____ number of electrons.

a. odd, odd, odd b. even, even, even c. odd, even, even d. odd, even, odd

e. even, odd, odd

31. As carbon-14 undergoes radioactive decay, its half-life

a. halves. b. doubles. c. is a constant. d. decreases at a constant rate.
e. increases at a constant rate.

32. A sample of radioactive material is allowed to remain undisturbed for a time equal to four half-life periods. What fraction of the original amount of material remains after that time?

a. 1/2 b. 1/4 c. 1/8 d. 1/16 e. 1/32

33. The decay constant is 0.0862/day for the spontaneous decay of iodine-131. If 10.0 grams of this isotope are allowed to stand for 16.1 days, how much of iodine-131 will remain?

a. 0.400 g b. 1.25 g c. 2.50 g d. 6.76 g e. 40.1 g

34. The half-life of sodium-24 is 15.0 hours. What is the value of the decay constant for sodium-24?

a. 21.6 hr^{-1} b. 7.50 hr^{-1} c. 10.4 hr^{-1} d. 0.0462 hr^{-1} e. 0.00216 hr^{-1}

35. What is the half-life of an isotope if the decay constant is 5.5/day?

a. 0.13 days b. 7.9 days c. 0.36 days d. 2.8 days e. 5.5 days

36. What is the half-life of an isotope if the decay constant is 3.2/year?

a. 0.58 year b. 1.6 year c. 0.22 year d. 8.7 year e. 6.4 year

37. The half-life of $^{14}_{6}C$ is 5730 years and its rate constant is 1.75 x 10^{-4}/year. If a tree dies and lies undisturbed for 18,400 years, what percentage of the $^{14}_{6}C$ remains?

a. 19.2% b. 17.4% c. 10.8% d. 8.43% e. 0.053%

38. The half-life of ^{90}Sr is 28 years. How long will it take for a sample of ^{90}Sr to be 85% decomposed?

a. 23 years b. 77 years c. 83 years d. 94 years e. 110 years

39. The half-life of ^{32}P is 14.3 days. How much of a 15.0 gram sample of ^{32}P will remain after 75 days?

a. 5.5 g b. 3.7 g c. 2.63 g d. 0.64 g e. 0.40 g

40. The half-life of iodine-131 is 8.0 days. If you have 25.0 grams of iodine-131, how much will remain after 40 days?

a. 20. g b. 5.0 g c. 3.1 g d. 0.78 g e. 0.039 g

41. The actual mass of oxygen-16 is 15.9949 g/mol. Calculate the binding energy of oxygen-16 based on the given information. Mass of a proton is 1.00783 g/mol. Mass of a neutron is 1.00867 g/mol. c = 3.00×10^8 m/s   1 J = 1 kg·m²/s²

a. 2.38×10^{13} J/mol   b. 2.38×10^8 J/mol   c. 1.23×10^{13} J/mol   d. 1.23×10^8 J/mol

e. 1.73×10^8 J/mol

42. Which of the following equations represent(s) a fusion reaction?

Reaction 1: $2\,H_2O \rightarrow 2\,{}^{2}_{1}H_2 + {}^{16}_{8}O_2$

Reaction 2: ${}^{2}_{1}H + {}^{3}_{1}H \rightarrow {}^{4}_{2}He + {}^{1}_{0}n$

Reaction 3: ${}^{235}_{92}U + {}^{1}_{0}n \rightarrow {}^{144}_{54}Xe + {}^{90}_{38}Sr + 2\,{}^{1}_{0}n$

a. Reaction 1   b. Reaction 2   c. Reaction 3   d. Reactions 1 and 2   e. Reactions 2 and 3

43. What role do cadmium rods play in nuclear reactors?

a. provide an alternative nuclear fuel

b. give off electrons for the initiation

c. absorb neutrons to control the rate of fission

d. fuse with ruthenium to initiate the reaction

e. serve as an inert matrix for the reaction

44. What is plasma?

a. the transient product of a fission reaction

b. unbound nuclei and electrons which may undergo fusion reactions easily

c. a liquid form of a radio active element

d. an unstable nucleus which can easily undergo fission

e. a group of positrons

45. Which elements are in the greatest abundance in the sun?

a. hydrogen and oxygen   b. hydrogen and helium   c. helium and tritium

d. helium and deuterium   e. deuterium and tritium

46. Artificial (synthetic) radioactive isotopes of many elements used in medical treatments

1. have short half-lives.
2. are made in fusion reactions.
3. emit γ radiation.

a. 1 only b. 2 only c. 3 only d. 1 and 2 only e. 1 and 3 only

47. In what year was the first nuclear bomb used as a weapon?

a. 1925 b. 1945 c. 1955 d. 1965 e. 1975

48. Of these chemists, which one did not play an active role in our current understanding of nuclear chemistry?

a. Bohr b. Meitner c. Einstein d. Boyle e. Fermi

49. The major problem in exploiting nuclear fusion as an energy source is

a. the high radioactivity of the reactants.
b. the high radioactivity of the products.
c. the instability of lithium deuteride.
d. the stability of lithium deuteride.
e. the high activation energy for nuclear fusion.

50. Enriched uranium is uranium that has a greater proportion of

a. lead-207. b. lead-208. c. uranium-235. d. uranium-236. e. uranium-238.

Chapter 24: Answers:

1. b	26. a
2. c	27. e
3. d	28. d
4. a	29. c
5. c	30. a
6. b	31. c
7. d	32. d
8. a	33. c
9. d	34. d
10. e	35. a
11. e	36. c
12. b	37. c
13. a	38. b
14. d	39. e
15. d	40. d
16. a	41. c
17. d	42. b
18. c	43. c
19. b	44. b
20. b	45. b
21. e	46. e
22. b	47. b
23. c	48. d
24. c	49. e
25. d	50. c